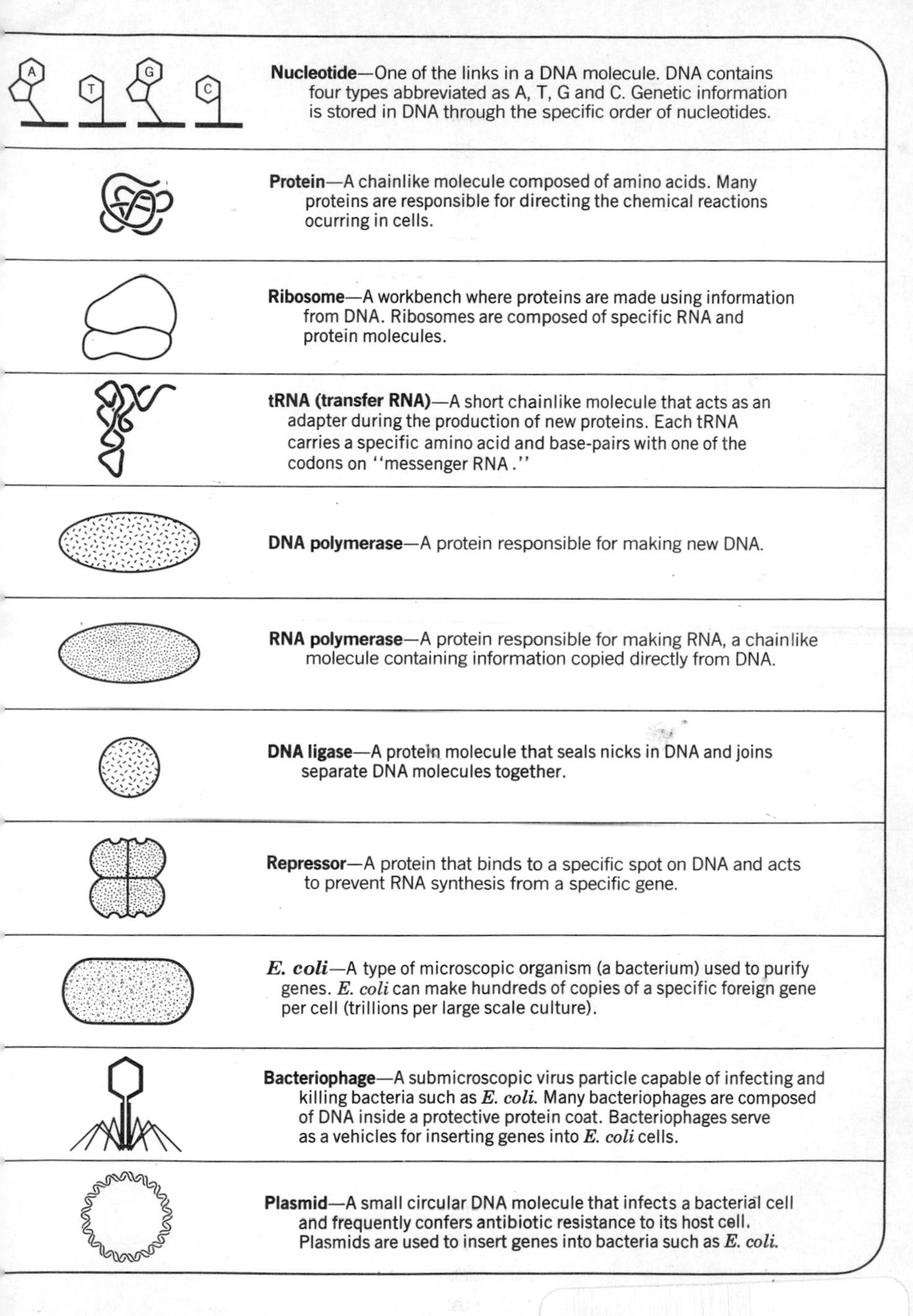

Nucleotide—One of the links in a DNA molecule. DNA contains four types abbreviated as A, T, G and C. Genetic information is stored in DNA through the specific order of nucleotides.

Protein—A chainlike molecule composed of amino acids. Many proteins are responsible for directing the chemical reactions ocurring in cells.

Ribosome—A workbench where proteins are made using information from DNA. Ribosomes are composed of specific RNA and protein molecules.

tRNA (transfer RNA)—A short chainlike molecule that acts as an adapter during the production of new proteins. Each tRNA carries a specific amino acid and base-pairs with one of the codons on ''messenger RNA.''

DNA polymerase—A protein responsible for making new DNA.

RNA polymerase—A protein responsible for making RNA, a chainlike molecule containing information copied directly from DNA.

DNA ligase—A protein molecule that seals nicks in DNA and joins separate DNA molecules together.

Repressor—A protein that binds to a specific spot on DNA and acts to prevent RNA synthesis from a specific gene.

E. coli—A type of microscopic organism (a bacterium) used to purify genes. *E. coli* can make hundreds of copies of a specific foreign gene per cell (trillions per large scale culture).

Bacteriophage—A submicroscopic virus particle capable of infecting and killing bacteria such as *E. coli.* Many bacteriophages are composed of DNA inside a protective protein coat. Bacteriophages serve as a vehicles for inserting genes into *E. coli* cells.

Plasmid—A small circular DNA molecule that infects a bacterial cell and frequently confers antibiotic resistance to its host cell. Plasmids are used to insert genes into bacteria such as *E. coli.*

Understanding DNA and Gene Cloning:
A Guide for the Curious

Understanding DNA and Gene Cloning
A Guide for the Curious

Karl Drlica
Public Health Research Institute
and New York University

John Wiley & Sons, Inc.
New York • Chichester • Brisbane • Toronto • Singapore

Acquisitions Editor	Sally Cheney
Copy Editing Manager	Deborah Herbert
Production Manager	Linda Muriello
Senior Production Supervisor	Savoula Amanatidis
Designer	Pete Noa
Cover Design	Ben Arrington
Illustration Coordinator	Sigmund Malinowdki
Manufacturing Manager	Lorraine Fumoso

Recognizing the importance of preserving what has been written, it is a policy of John Wiley & Sons, Inc. to have books of enduring value published in the United States printed on acid-free paper, and we exert our best efforts to that end.

Library of Congress Cataloging in Publication Data:

Drlica, Karl.
Understanding DNA and gene cloning : a guide for the curious / Karl Drlica.—[2nd ed.]
p. cm.
Includes index.
ISBN 0-471-62225-7 (paper)
1. Molecular cloning. 2. Recombinant DNA. 3. Genetic engineering. I. Title.
QH442.2.D75 1992 91-26076
574.87′328—dc20 CIP

Printed in the United States of America

10 9 8 7 6 5 4

Printed and bound by Courier Companies, Inc.

To Ilene, for many years of support

PREFACE

Gene cloning technologies continue to spur advances in many biological disciplines, and an update of *Understanding DNA* is long overdue. As with the first edition, my goal is to explain the fundamental principles of DNA biology at a level that does not require a knowledge of chemistry. Thus the discussion starts at a much more elementary level than is commonly found in publications such as *Scientific American* and *The New York Times*. It then builds on the basics to introduce the reader to some of the more sophisticated concepts of molecular biology by using drawings and in some cases analogies.

Understanding DNA was written for college students. In junior colleges the readers have often been potential biology majors, whereas in elite private universities they have tended to be nonscience majors. But it has been gratifying to see *Understanding DNA* reach a much wider audience, one that has included investment bankers, mechanical engineers, medicinal chemists, and precocious high school students. The book even caused a Ph.D. physiologist to begin using gene cloning in a research project involving kidney function.

Several elements have been added or expanded to make the second edition a more comprehensive teaching tool. One is the addition of questions for discussion. Some of these questions have specific answers to reinforce a point; others are open-ended to stimulate additional reading. Many of the questions also introduce information that would dilute the main themes if this information were included in the primary text. Another teaching aid is the glossary, which has been expanded. Vocabulary is a key aspect of learning molecular biology, and the reader should expect to refer frequently to the glossary. As an instructor I found that administering simple glossary quizzes early in a course overcomes some of the vocabulary barriers. A third aid is the list of additional readings, which has also been expanded. Most of the new entries are from *Scientific American* because the articles are of high quality, they are at the appropriate level when used in conjunction with *Understanding DNA,* and *Scientific American* is readily available.

Several important topics have been added to the second edition. Among them is a discussion of retroviruses, viruses that cause cancer and AIDS. AIDS is a relatively new disease that is affecting us all; even those not exposed to the virus itself are seeing personal relationships become more circumspect, are being troubled by controversial educational programs, and are witnessing medical institutions and insurance companies alter their services to cope with the expense of the disease. Another addition is a brief description of monoclonal antibodies, immunological tools that greatly expand our ability to study the biology of specific proteins. When combined with gene cloning strategies, monoclonal antibody techniques provide incredibly precise methods for learning how biological information is stored, transmitted, and used by cells. Other topics added are gene amplification (polymerase chain reaction), Southern hybridization, ribozymes, and the Ti plasmid.

I would like to thank a number of people for helping with the second edition. John Balbalas deserves special acknowledgement for the illustrations. Deborah Everett, Barry Kreiswirth, Ellen Murphy, Abraham Pinter, Steve Projan, Todd Steck, and Shermaine Tilley provided many excellent ideas and valuable criticisms; John Kornblum and Brenda Griffing painstakingly edited the manuscript and turned up many flaws that had escaped me in the first edition.

Karl Drlica

PREFACE TO THE FIRST EDITION

An explosion of knowledge is shaking the science of biology, an explosion that will soon touch the life of each one of us. At its center is chemical information—information that our cells use, store, and pass on to subsequent generations. With this new knowledge comes the ability to manipulate chemical information, the ability to restructure the molecules that program living cells. Already this new technology is being used to solve problems in diverse areas such as waste disposal, synthesis of drugs, treatment of cancer, plant breeding, and diagnosis of human diseases. The new biology is also telling us how the chemicals in our bodies function; we may soon be programming ourselves and writing our own biological future. When this happens, each of us will be confronted with a new set of personal and political choices. Some of these difficult and controversial decisions are already upon us, and the choices will not get easier. Informed decisions require an understanding of molecular biology and recombinant DNA technology; this book is intended to provide that understanding.

Molecular biology is a science of complex ideas supported by test tube experiments with molecules. Consequently, the science has remained largely inaccessible to those without a knowledge of chemistry. I hope to change that situation—this book requires the reader to have little or no background in chemistry. Chemical processes and molecular structures are described by means of analogies using terms familiar to nonscientists. Technical terms have been kept to a minimum; where they must be introduced, they are accompanied by definitions. In addition, a glossary has been provided for easy reference; items in the glossary are in boldface type the first time they appear in the text.

It is also my intent to provide a sense of how informational molecules are manipulated experimentally. Integration of these details should help remove the mystery from gene cloning and expose the elegance and simplicity of the technology. I hope that this brief introduction to gene cloning will help you enjoy and appreciate the science of molecular biology for the art form that it is.

A number of people have helped me in this endeavor, and they deserve most of the credit for making this book readable. I especially thank Lynne Angerer, Betty Bonham, Tom Caraco, Cheryl Cicero, Lisa Dimitsopulos, Dianne Drlica, Karen Drlica, Rob Franco, Claire Gavin, Ed Goldstein, Brenda Griffing, George Hoch, Hiroko Holtfretter, Johannes Holtfretter, Lasse Lindahl, Stephen Manes, Bill Muchmore, Pat Pattison, Donna Riley, Peter Rowley, Ron Smith, Franklin Stahl, Todd Steck, Ilene Wagner, William Wasserman, Grace Wever, Bill Wishart, and Janice Zengel. I also acknowledge Alvin J. Clark and Henry Sobel for technical information and Ron Sapolsky for artistic insights used in early versions of the manuscript. Fred Corey and his staff at John Wiley & Sons provided excellent editorial assistance. The illustrations are the creative work of John Balbalis; where appropriate he has attempted to provide a sense of relative scale among the elements involved.

Karl Drlica

INTRODUCTION

These days, all of us are constantly exposed to the on-going revolution in biological knowledge. One's daily newspaper is likely to contain an announcement of an advance in our understanding of a disease such as cancer and AIDS, or an article about how the rapid increase in our detailed knowledge of thousands of human genes is leading to new forms of disease diagnosis and therapy. Yet I have been told by members of the press that most of their readers know almost nothing about modern biology—and that most of them are even unclear about the difference between a chromosome, a gene, and a DNA molecule. If so, then something is drastically wrong.

As was pointed out more than sixty years ago, "the key to every biological problem must finally be sought in the cell, for every living organism is, or at sometime has been, a cell" (E. B. Wilson, "The Cell in Development and Heredity," 1925). Each of us, for example, originates as a single cell (formed from the fusion of one sperm cell and one egg cell), which grows and divides until it produces a highly organized cooperative of more than 10 million million cells—the adult human. A typical cell is so small that it would take 10,000 of them to cover the head of a pin. But, the relatively simple behaviors of individual cells have, in aggregate, a surprising power to explain even sophisticated properties of multicellular organisms—such as the memory stored in

---

This introduction is an extension of ideas originally expressed by Bruce Alberts in *The American Zoologist,* 1989.

the nerve networks in our brains, or the growth and patterning of a developing embryo. Those of us who understand nothing at all about cells, like those of us who know nothing at all of poetry, are missing out on one of the great pleasures of human existence—the search to understand ourselves.

But what exactly is a cell? Originally defined as the smallest unit of an organism that is alive, to biologists today a cell is nothing more (or, much better, nothing less) than a special collection of complex molecules, enclosed by a membrane and having the very special ability to reproduce itself from the much simpler molecules available in its surroundings. Speaking as a chemist, a cell is a self-replicating collection of catalysts—most of which are proteins. We know how this works in principle, but a complete understanding of a living cell will require that we know every reaction that occurs in it, so we can see how each component contributes to the self-replication of the entire unit. In time, this knowledge will come—although even the simplest known cell, the tiny bacterium known as *Mycoplasma*, is estimated to contain a total of 40,000 protein molecules, of about 600 different kinds.

Cells and organisms are very complex. But, because they have evolved to this complexity by a repeated process of DNA sequence duplication and divergence, each cell is composed of parts that are closely related to other parts of the same cell in their structure and function. This fact greatly simplifies the task of understanding the tens of thousands of proteins that make a human being. For the same reason, there is a surprising uniformity among living things. We know from DNA sequence analyses that plants and higher animals are closely related, not only to each other, but to relatively simple single-celled organisms such as yeasts. Cells are so similar in their structure and function that many of their proteins can be interchanged from one organism to another. For example, yeast cells share with human cells many of the central molecules that regulate their cell cycle, and several of the human proteins will substitute in the yeast cell for their yeast equivalents!

Scientists who have devoted their lives to studying cells view the cell as a large and elegant puzzle. Each biological macromolecule (protein, nucleic acid, or polysacharide) that is discovered and studied in detail represents a small piece of the puzzle, which will only be satisfactorily understood when it has been adequately connected to all of the other "pieces" in the cell with which it interacts. Ten years ago, our total amount of information about cells was so small that most of the interconnections between the known pieces were missing. In the last few years, we seem to have reached the point where enough of the puzzle has been filled in that each new piece analyzed (most often a new protein) can often be connected to several others to provide some new insight. In terms of the puzzle analogy, we are still far from seeing the final picture, but we can often glimpse part of a tree, or recognize a familiar face in an otherwise chaotic jumble of partial information about the cell.

Much of the present excitement in biology stems from the feeling that we are now starting to know enough about cells to derive the type of connections that make conceptual sense of what seemed previously to be only an inexplicable muddle of facts. Moreover, by detailed comparisons of the components in the cells of *different* present-day organisms, we can hope to solve an even bigger puzzle than that of the workings of the cells themselves: what is the exact pathway by which living organisms evolved on the earth? This "megapuzzle" represents perhaps the ultimate intellectual challenge for future biologists.

The modern emphasis on explaining biological phenomena in terms of the behavior of molecules reflects the belief of today's biologists that the tools are in hand to achieve such a detailed, mechanistic explanation. This book describes the most revolutionary of these tools: gene cloning and the accompanying recombinant DNA technology. These methods, unforeseen as little as 20 years ago, have made it possible to answer almost any question about the cell, given sufficient effort. How quickly biology has changed! When I worked as a graduate student in the early 1960s, the cell seemed incomprehensibly complex. Most importantly, there seemed to be no obvious way of deciphering this complexity. Most of the tens of thousands of different protein molecules in a higher eukaryotic cell were known to be present in such small amounts that it appeared impossible to ever know their structure. As lucidly explained in this book, this situation has entirely changed. With gene cloning and the ability to manipulate the cloned genes so as to produce any gene product, every protein in the cell is potentially accessible in virtually unlimited amounts. Like the first settlers to arrive in California for the Gold Rush of 1849, today's biologist faces an easy harvest of riches. For the next twenty years or so, one need not be especially clever or wise to make a major contribution to biology. With luck, even the random cloning of a new gene—which requires relatively little skill and no insight—can turn out to be exciting.

That cells exist at all is a marvel. To speak about them as "simply a self-replicating collection of catalysts" in no way reduces the beauty and wonder of the living state. If Earth were to be visited by a being from outer space, this being would undoubtedly find even the simplest of the living cells far more fascinating than any human-made object. That our children are largely bored with cells and the rest of biology—but fascinated by consumer electronics and automobiles—is a great tragedy, and it reflects how far we have to go in changing how science is taught to the general public. If this book by Karl Drlica can make a contribution to bringing an appreciation of the beauty of cells to others, the world will be much richer for it.

Bruce M. Alberts
American Cancer Research Professor of Biochemistry
University of California, San Francisco

INTRODUCTION TO THE FIRST EDITION

In thinking about the course of human events, it has often occurred to me that they very much resemble the course of a river. A river meanders, gathers small streams, widens, deepens, and may even split into smaller rivers that go their separate ways. On occasion, rivers merge, a confluence that creates a mightier river. In the same sense, the extraordinary developments in genetic chemistry are part of an even more profound development in medical science, a change that is truly revolutionary. It is the confluence of the many discrete and previously unrelated medical science subjects into a single, unified discipline. Anatomy, physiology, biochemistry, microbiology, immunology, and genetics have now merged and are expressed in a common language of chemistry. By reducing structures and systems to molecular terms, all aspects of body form and function blend into a logical framework. Universities still maintain departmental lines to define administrative boundaries, but they are now meaningless in the pursuit of new knowledge.

The remarkable confluence of medical science first appeared in the genius of Louis Pasteur. More than any individual or school, he established medicine as a science and gave it the form we recognize today. Pasteur was trained as a chemist. His first exploit as a very young man was to show that two samples of tartaric acid of identical chemical composition differed physically because the molecules were mirror images of each other. Pasteur's "germ theory of disease" bore the stamp of his chemical background. He tried to reduce a problem of disease to elementary components. His experimental approach was to purify the causative agents to homogeneity and recreate the disease with the isolated pure form of the agent. From this he created and practiced the disciplines of microbiology and immunology. It might surprise many microbiologists and immunologists today to find that in 1911 the *Encylopaedia Britannica* described Pasteur as a French *chemist,* the acknowledged head of the greatest *chemical* movement of his time.

Pasteur's scientific career had a flaw. Having established that the yeast cell is responsible for the conversion of sugar to alcohol, he tried to extract from the yeast cell the juices that would do the same thing. In this he failed and so concluded that nothing short of a living cell could possibly carry out this very complex chemical reaction. Pasteur's self-confidence, persuasiveness, and influence were so great that attempts by others to obtain alcoholic fermentation in a cell-free system were severely discouraged. And so, cellular vitalism became firmly rooted, and the advent of modern biochemistry was delayed for 30 years.

Only at the turn of this century did Eduard Büchner in Munich accidentally discover fermentation by disrupted yeast cells. In employing sugar as a preservative for yeast extracts, he observed a strange frothing. He had the insight to identify carbon dioxide as the gas and ethanol as the product of sugar degradation by the yeast juice. It was Pasteur's poor fortune that his extracts of Parisian yeast were deficient in sucrase, the enzyme that initiates the pathway of sugar metabolism. Luckily for Büchner, adequate amounts of the enzyme survived in his extracts from Munich yeast.

The reactions by which a yeast cell converts sugar to ethanol and carbon dioxide could then be isolated and analyzed in detail. In all, a dozen discrete, complex molecular rearrangements, condensations, and scissions are needed to achieve the fermentation of sugar to alcohol. Each of these reactions is catalyzed by an elaborate protein, an enzyme, designed to carry out the singular operation. The enzyme increases the rate of the reaction by a million- or trillionfold and gives it a unique direction among the many potential fates to which the molecule is susceptible.

These revelations of alcoholic fermentation in yeast provided the methods and confidence for the investigation of a comparable question. How does a muscle derive energy from sugar to do its work? When that mystery was unraveled, the plot and most of the characters in the muscle story incredibly proved to be the same as in yeast. There is, of course, one deviation. In muscle at the final stage, lactic acid is produced instead of alcohol and carbon dioxide.

Reconstitution in the test tube of the yeast and muscle pathways of sugar combustion to generate usable energy set the stage for a generation of enzyme hunters in the 1940s and 1950s. My own attempts at synthesizing DNA with enzymes in a test tube were regarded by some as audacious. Reconstitution of the metabolism of fats as well as carbohydrates may be one thing, but the enzymatic synthesis of genetically precise DNA, thousands of times larger, must be quite another. Yet all I have done is follow in the classical traditions of biochemists of this century. It always seemed to me that a biochemist with a devotion to enzymes could, with sufficient effort, reconstitute in the test tube any metabolic event as well as the cell does it. In fact, the biochemist, freed from the cellular restraints of the concentrations of enzymes, substrates, ions, and metals, and with the license to introduce reagents that retard or drive a reaction, should do it even better.

As the disciplines of genetics, microbiology, and physiology reached more and more for chemical explanations, they began to coalesce with the biochemistry of the enzyme hunters. From this coalescence came molecular biology and genetic engineering. Narrowing our focus to the molecular biology of DNA, I would cite several diverse origins. One origin is in medical science. In 1944 Oswald Avery, in his lifelong and relentless search for control of pneumococcal pneumonia, became the first to show that DNA is the molecule in which genetic information is stored. A second origin of molecular biology is in microbial genetics. In the late 1940s and early 1950s microbiologists, some of them renegade physicists, chose the biology of the small bacterial viruses, the bacteriophages, to elucidate the functions of the major biomolecules: DNA, RNA, and proteins. At about the same time a third origin of molecular biology arose as the structural chemistry of these biomolecules became highly refined. Analysis of the X-ray diffraction patterns of proteins revealed their three-dimensional structures; the DNA patterns gave us the double helix and a major insight into its replication and function. A fourth origin of molecular biology is in biochemistry, the enzymology, analysis, and synthesis of nucleic acids. The biochemist provided access to the nucleases that cut and dissassemble DNA into its genes and constituent building blocks, the polymerases that reassemble them, and the ligases that link DNA chains into genes and the genes into chromosomes; these are the reagents that have made genetic engineering possible. In the cell, these enzyme actions replicate, repair, and rearrange the genes and chromosomes.

Molecular biologists practice chemistry without calling it such. They identify and isolate genes from huge chromosomes, often only one part in millions or billions, and then they amplify that part by even larger magnitudes using microbial cloning procedures. They map human chromosomes, analyze their composition, isolate their components, redesign their genetic arrangement, and produce them in bacterial factories on a massive industrial scale. New species are created at will. Not even the boldest among us dreamed of this chemistry 10 years ago. I generally underestimated how permissive *E. coli* would be at accepting and expressing foreign genes. As the effects of a more profound grasp of chromosome structure and function become manifest, the impact on medicine and industry will prove to be far greater even than extrapolations from the current successes in the mass production by genetic engineering of rare hormones, vaccines, interferons, and enzymes.

Since the role of basic research is not always apparent to the general public, I would like to make another historical comment. Genetic engineering is solely an outgrowth of basic research. It was never planned, nor was it even clearly anticipated. Many of the procedures were discovered as unanticipated by-products of experiments designed to satisfy someone's curiosity about nature. For example, the analyses and rearrangements of DNA that form the drama of genetic engineering depend largely on a select cast of enzymes. Yet these actors were neither discovered nor created to fill these roles. Some of these enzymes, uncovered in my laboratory, came from a curiosity about the mechanisms of DNA replication. In these explorations, sponsored by the National Institutes of Health and the National Science Foundation for more than 25 years at a total cost of several million dollars, I neither anticipated nor promised their industrial application. Nor did any of my colleagues with comparable, federally funded projects. Thus, the multibillion dollar industry projected by Wall Street is entirely a product of the knowledge and opportunities gained from the pursuit of "irrelevant," basic research in universities, research made possible by the investment of many hundreds of millions of dollars by federal agencies over more than two decades.

As we retrace the flow of knowledge, we see that the first two decades of twentieth-century medical science were dominated by the microbe hunters. Their place in the spotlight was superseded for two decades by the vitamin hunters. They in turn were succeeded by the enzyme hunters in the 1940s and 1950s. For the past two decades the gene hunters have been in fashion. To whom the remaining years of our century will belong is uncertain. The neurobiologists, call them the head hunters, may very well claim it. If so, we will again see how chemistry is the fundamental language. Although brain chemistry may be novel and very complex, it is expressed in the familiar elements of carbon, nitrogen, oxygen, and hydrogen, of phosphorus and sulfur that constitute the rest of the body. Brain cells have the same DNA that all cells do; the basic enzyme patterns are those found elsewhere in the body. It is now known that hormones once thought to be unique to the brain are produced in the gut, ovary, and other tissues. The form and function of the brain and nervous system must ultimately be explained in terms of chemistry. The repeated failures of science to analyze social, economic, and political systems should not discourage us from pursuing the idea that individual human behavior, at least, can be explained by physical laws.

I sense in the future a better awareness that life can be described in rational terms and a furtherance of chemical language to express it. For chemistry is a truly international language. It links the physical and biological sciences, the atmospheric and earth sciences, the medical and agricultural sciences. The chemical language is a rich and fascinating language that creates images of great aesthetic beauty. I see the language of chemistry taught and used for the clearest statements about our individual selves, our environment, and our society. Such visions excite me. I hope you share them. They give us courage to face the future.

Arthur Kornberg
Stanford University

CONTENTS

CHAPTER ONE

PREVIEW

DNA, Gene Cloning, and Public Safety

Overview

Information governing the characteristics of all organisms is stored in long, thin molecules of deoxyribonucleic acid (DNA). DNA molecules contain regions (genes) that specify the structure of other molecules called proteins. Protein molecules in turn control cellular chemistry and contribute to cell structure.

Biologists can now obtain large amounts of specific regions of DNA. The general strategy involves first cutting the DNA into small fragments. The fragments are then moved into single-celled organisms, often bacteria or yeast. Conditions are set up so the fragments become a permanent part of the organism, and pure cultures of bacteria or yeast containing a particular DNA fragment are then obtained. The fragments are removed and used for further study, for detecting disease, for changing the cellular chemistry of another organism, and for producing large quantities of specific proteins of medical or industrial value.

Since microorganisms receiving DNA fragments are being changed in unknown ways, gene cloning experiments were originally viewed as **potentially** dangerous. Regulations were therefore established to minimize the potential health hazards. Many types of recombinant DNA have now been constructed and studied; no harmful effects have been observed, and almost all biologists now consider the cloning of genes to be safe.

INTRODUCTION

In a general sense biologists have solved the riddle of heredity, the question of why offspring resemble their parents. The explanation of heredity lies in the chemical behavior of **submicroscopic** structures called **molecules**. (All boldface words are included in the glossary.) At the center of this new understanding is a giant molecule called **deoxyribonucleic acid (DNA)**. This book is about DNA, the chemical that specifies features such as eye color and blood type. DNA in our cells influences all of our physical characteristics, as is true for every living organism on earth.

This book is also about genetic engineering, recombinant DNA, and gene cloning. In particular, it is about how gene cloning works and about some of the things we have learned from it. The goal of this first chapter is to provide a framework for the rest of the book. A discussion of DNA structure and function is started, some comments are made about fundamental features of atoms and molecules, **gene cloning** is introduced, and safety issues are briefly discussed. The chapter concludes by outlining the types of information presented in subsequent chapters.

The basic unit of life is the **cell**, an organized set of chemical reactions bounded by a membrane and capable of self-perpetuation. Our bodies are collections of trillions of cells working together, with each cell having its own identity and function. For example, liver cells cluster together to form livers, and skin cells attach to each other to cover our bodies. With few exceptions, every cell contains all the information required for an independent existence; indeed, under the right conditions human cells can be removed from the body and grown in laboratory dishes.

The information necessary to control the chemistry of the cell (i.e., the chemistry of life) is stored in the long, thin fiber called DNA. DNA fibers are found in every cell except mature red blood cells, and they dictate how a particular cell behaves. Thus, DNA controls our body chemistry by controlling the chemistry of each of our cells.

Isolated DNA looks like a tangled mass of string (Figure 1-1). Our cells, which are generally less than a **millimeter** long, contain about 2 meters of DNA specially packaged to fit inside. DNA can be bent, wrapped, looped, twisted, and even tied into knots. Many DNA molecules are circles, which are sometimes found interlinked like a magician's rings. In terms of three-dimensional structure, DNA is very

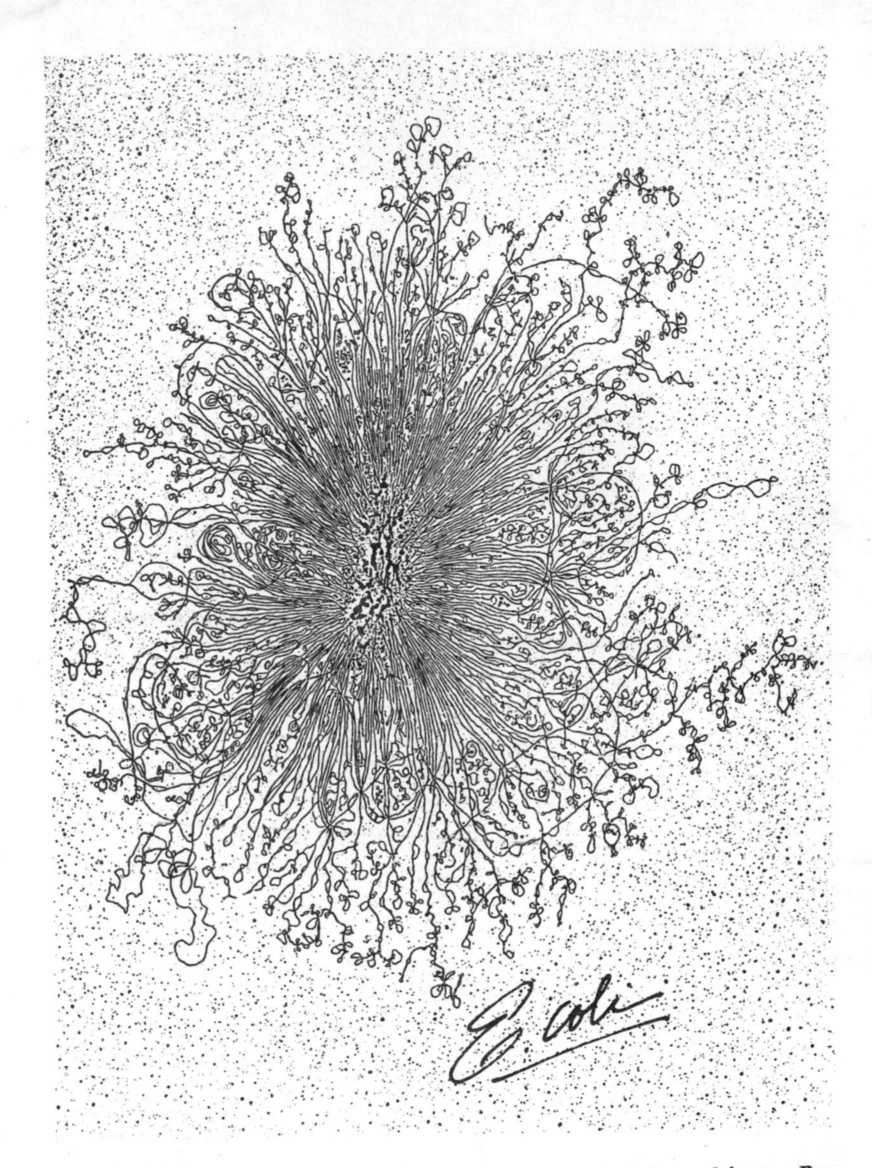

Figure 1-1 Electron Micrograph of a DNA Molecule Released from a Bacterium. The long, threadlike material is DNA, which is about one millimeter long or about 1000 times the length of the bacterium from which it was taken. The molecular details of how this DNA is compacted and packaged to fit in the cell are not yet understood. The electron micrograph is of a purified, surface-spread *E. coli* chromosome prepared by Ruth Kavenoff and Brian Bowen. The line under the *E. coli* signature represents 2.5 micrometers. Copyright © with all rights reserved by Designer-Genes Posters, Postcards, T-shirts, etc., P.O. Box 100, Del Mar, CA 92014.

flexible. But in terms of information content, DNA is quite rigid, for the same information must pass from generation to generation with little change.

One of the goals of this book is to explain how information is stored in DNA, how it is used, and how it is reproduced. For now, the important concept is that distinct regions of DNA contain distinct bits of information. The specific regions of information are called **genes**. In some ways DNA is similar to motion picture film. Like film, DNA is subdivided into "frames" that make sense when seen in the correct order. In DNA the "frames" correspond to the letters in the genetic code, which are described in Chapter 3. When a number of frames or genetic letters are organized into a specific combination, they create a scene in the case of film and a gene in the case of DNA (Figure 1-2).

Information in genes is used primarily for the manufacture of **proteins**. Proteins are chainlike molecules that fold in a precise way to form specific structures. Some proteins contribute to the architecture

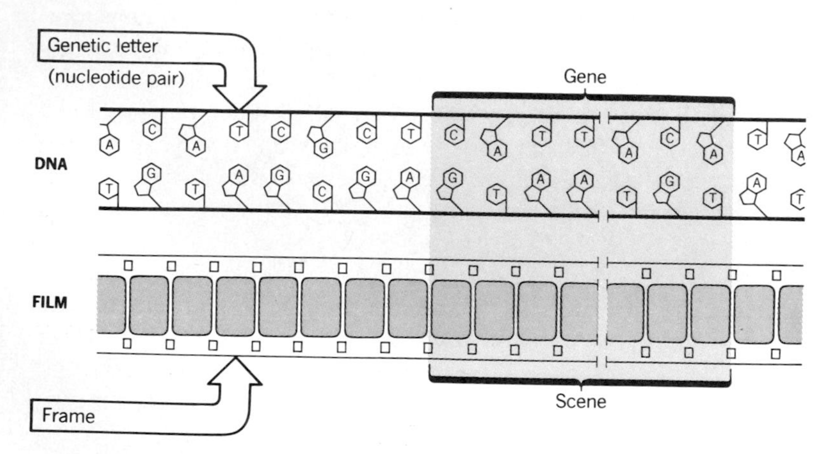

Figure 1-2 Comparison of DNA and Motion Picture Film. The frames of movie film correspond to genetic letters (**nucleotide pairs**) in DNA. When properly organized, the frames form a scene in film, and the genetic letters form a gene in DNA. DNA contains many genes, and each one stores information affecting a chemical process that occurs in living cells. Genes are generally hundreds to thousands of nucleotide pairs long. For illustrative purposes only a part of a gene is shown in the figure; slashes have been drawn through the DNA and film to indicate that many nucleotide pairs and frames have been omitted.

of the cell, while others directly control cell chemistry. Occasionally we can easily see the effects of particular genes and proteins; for example, a small group of genes is responsible for determining eye color. It is the specific information in the DNA, in the genes, that makes human beings different from honey bees or fir trees. Information in your DNA makes you different from anyone else on earth—unless you have an identical twin.

ATOMS AND MOLECULES

Each DNA fiber is a **molecule**, a group of **atoms** joined together to form a distinct unit. Several points are important for understanding discussions of atoms and molecules. First, all forms of matter are composed of submicroscopic particles called atoms. Second, molecules are very specific combinations of atoms, and the combinations have distinctive properties. Third, atoms and molecules can join with each other or with single atoms to form new molecules—that is, new combinations of atoms whose properties differ from those of the starting materials. Such interactions are called **chemical reactions**. For example, hydrogen gas is two atoms of hydrogen, and oxygen gas is two atoms of oxygen. If you combine an oxygen atom with two hydrogen atoms, you get water, a molecule very different from either oxygen or hydrogen. Fourth, the number of **elements**, or atoms of different kinds, is small (about 100). The number of types found in living material is still smaller (the major ones are listed, along with their common abbreviations, in Figure 1-3). The small number greatly simplifies the process of converting one substance to another. A fifth point is that large molecules such as DNA are composed of many smaller groups of atoms. By understanding the properties of the smaller groups, it is possible to predict how they will behave when joined together. Thus it is not necessary to know the precise position of every atom in DNA to understand how the giant molecule acts.

To discuss the properties of molecules, it is often useful to draw pictures in which the relative positions of the component atoms are specified. These pictures, or structural formulas, help explain the properties of molecules. An example of a sugar molecule is shown in Figure 1-4. Frequently these pictures will be abbreviated, and often some of the atoms will be omitted for clarity (compare the different ways of representing sugar structure illustrated in Figure 1-4).

Name	Symbol	Atomic weight	Properties of pure element
Hydrogen	H	1.01	Light, colorless gas
Carbon	C	12.01	Hard solid (diamond, graphite)
Nitrogen	N	14.01	Colorless gas
Oxygen	O	16.00	Colorless gas
Fluorine	F	19.00	Pale greenish gas
Sodium	Na	23.00	Reactive silver metal
Magnesium	Mg	24.31	Light, silvery metal
Phosphorus	P	30.97	White, red, or yellow nonmetal
Sulfur	S	32.06	Yellow solid
Chlorine	Cl	35.45	Yellow-green gas
Potassium	K	39.10	Light, silver-white metal
Calcium	Ca	40.08	Soft, silvery metal
Manganese	Mn	54.94	Hard, brittle metal
Iron	Fe	55.85	Silvery grey metal
Copper	Cu	63.54	Malleable reddish metal
Zinc	Zn	65.37	Blueish-white metal
Selenium	Se	78.96	Red or grey semimetal
Molybdenum	Mo	95.94	Tough, silvery metal
Iodine	I	126.90	Violet crystalline nonmetal

Figure 1-3 Atoms (Elements) Commonly Found in Living Material. As with all matter, living organisms are composed of specific combinations of atoms. The chemical symbol and the relative size (atomic weight) are listed for each of the elements (i.e., atom types) that are found in organisms. Adapted from E. O. Wilson, et al., *Life,* Sinauer Assoc., Sunderland, Mass.

When atoms form a molecule, they are held together by forces called chemical bonds. Chemical bonding is generally discussed in terms of the structure of atoms. Atoms are composed of a nucleus surrounded by a cloud of electrons that circle the nucleus much as the planets orbit the sun. Atoms differ in nuclear size and electron number; the number and arrangement of electrons determines bonding patterns. Thus bonding patterns are characteristic of the specific atoms involved in an interaction. For example, carbon almost always forms four bonds to other atoms, and hydrogen forms only one (see Figure 1-4*a*). Some types of chemical bond are very strong, such as those formed when two atoms share an electron (such bonds are represented by the lines

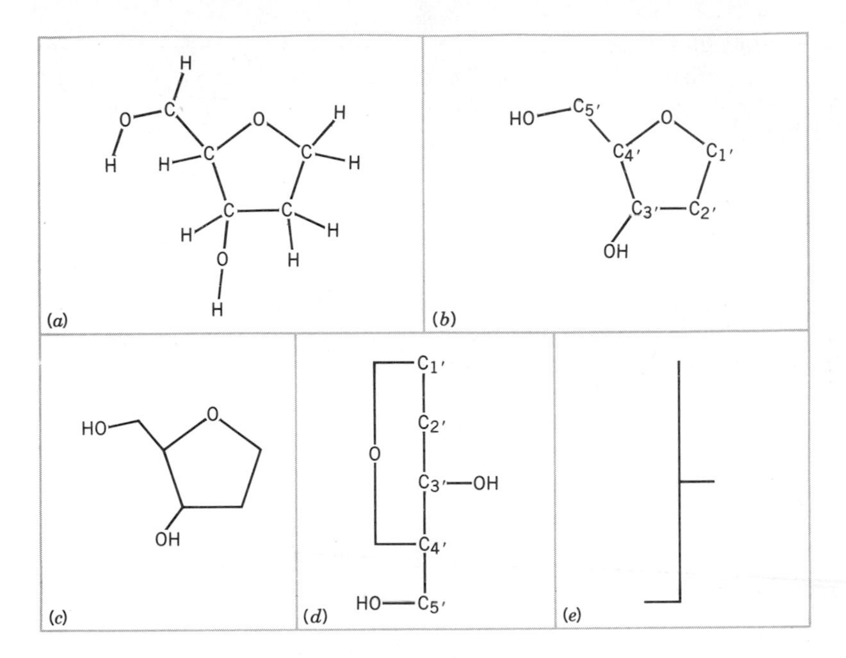

Figure 1-4 Symbolic Representations of a Molecule. The arrangement of atoms in a molecule can be illustrated in a number of ways. Frequently atoms are deleted from a diagram for clarity. Sometimes little attention is paid to the precise spatial orientation of the atoms (bond angles and lengths) because such details would obscure the point of the drawing. Here the arrangement of atoms in the sugar called deoxyribose is depicted in four ways. None of these includes information about spatial orientation. **(a)** All of the atoms are shown; the lines between the atoms represent the chemical bonds that hold the atoms together. **(b)** The hydrogens, except for the two attached to oxygen, have been removed from the diagram for clarity. **(c)** The structure is the same as in (b), but the carbon atoms are not explicitly indicated. **(d)** The carbon–oxygen ring in (b) is redrawn in a way that is useful for showing directionality in DNA, a point that is developed in later chapters. The carbon atoms are numbered in (b) and (d) to help relate the two diagrams. **(e)** The atoms have been eliminated from the structure in (d).

between the atoms in Figure 1-4). The strong bonds keep the atoms of each molecule together; they make each molecule a discrete physical entity. Other bonds are much weaker. These include the attractive forces that result from atoms having different electrical charges. Such bonds are easily broken. However, if many weak bonds form between two molecules, they can collectively be quite strong. Weak bonds are often responsible for bringing distant regions of a large molecule close or for holding two different molecules together. For two molecules to be held together, the molecules usually have to fit tightly (for example, see Figure 3-5*a*; the dotted lines represent weak bonds). The weak bonds allow biological molecules to recognize each other, to come together, and to separate when the conditions change. Thus one can think of DNA as a molecule that contains information specified by the arrangement of its atoms, held together by strong bonds. Weak bonds provide the forces that hold DNA strands together and allow DNA to interact physically with other molecules.

GENE CLONING

Operationally, gene cloning consists of performing a series of biochemical manipulations. It is analogous to baking a cake, and recipes are available for each process. In general, to clone a region of DNA one must first cut the DNA into specific pieces. The cutting tools usually produce so many different types of DNA fragment that individual types cannot be separated in a straightforward way. Instead, one-celled microorganisms are used to carry out the separation. Large numbers (billions) of microorganisms are mixed with the fragmented DNA in such a way that the DNA moves inside the microorganisms and takes up permanent residence there. Conditions are adjusted so each microorganism receives no more than one piece of DNA. The microorganisms are then scattered onto a solid surface. There they grow and reproduce; each forms a small cluster. The biologist then tests the clusters to determine which one contains the DNA fragment of interest. That cluster of cells is saved and allowed to reproduce many more times. Finally the cells are broken open, and the DNA fragment of interest is removed.

Cloned genes can be used to change the characteristics of organisms; thus we can expect gene cloning to have a major impact on agriculture as new breeds of plants and animals are developed. In the area of medicine we can expect gene therapy to help people suffering from genetic diseases, and genetic screening may eventually help some people adjust their behavior to avoid certain activities that, because of individual genetic makeup, constitute high-risk behavior.

Cloned DNA can also be used to produce medically important proteins. Generally a protein that controls a chemical reaction is very specific to that reaction. Consequently, if a person produces a defective protein, it is sometimes possible to replace this substance by injection without interfering directly with the other chemical reactions of the body. Some diabetics, who fail to produce sufficient quantities of the protein called **insulin**, are unable to properly control their **sugar metabolism**; consequently, these patients must take daily injections of insulin. Before the development of gene cloning, insulin could be obtained only by an expensive process of extracting the protein from hog pancreas; but now, through gene cloning techniques, human insulin genes have been placed in bacteria. Here the genes are expressed; that is, insulin is made inside bacteria. Thus, large quantities of insulin are now produced by bacteria, and it is easier to obtain insulin from bacteria than from pancreas tissue. Moreover, the engineered bacteria produce human insulin, an important feature for diabetics who have become allergic to hog insulin.

THE SAFETY CONTROVERSY

Shortly after the first gene cloning experiments were completed in the early 1970s, scientists realized that this type of genetic manipulation might pose health hazards. No danger had been demonstrated, but one could imagine a number of scenarios frightening enough to make good science fiction copy. Suppose, for example, that the gene containing the information for botulism **toxin** were placed inside a harmless bacterium and that large numbers of this new, toxin-producing bacterium were accidentally released into the environment. A few of these organisms might find their way into the digestive tracts of humans. There they would multiply, for the human digestive tract is one of the normal habitats for the type of bacterium most commonly used in gene cloning experiments (but not for the organism that normally produces botulism toxin). If the botulism toxin were produced by the "engineered" bacteria, anyone infected with these bacteria would soon die. Since little was known about the ecological relationships between common laboratory bacteria and man, there was no way to determine whether such a scenario could be realized. Thus it seemed prudent to use recombinant DNA technology with caution.

The scientists who developed gene cloning recognized the need for precautions. They tried to control the use of the **cloning vehicles**, the biological tools used to transfer genes, but it soon became apparent that they would be unable to do this by themselves. Gene cloning involves straightforward laboratory procedures, and the discoverers knew that soon hundreds of scientists all over the world would be conducting experiments with recombinant DNA. To be effective, the control system would require much more muscle than a handful of scientists could flex.

In the United States, a regulatory system was set up by the National Institutes of Health (NIH), the federal agency responsible for funding most recombinant DNA research. A set of guidelines was established that focused on containing recombinant organisms inside laboratories. In some cases the investigators could use standard laboratory techniques, whereas in others the experiments had to be conducted in specially constructed rooms isolated from the environment. Still other experiments were simply disallowed. The responsibility for compliance rested on each institution involved in recombinant DNA research. At stake was not only the health of the local community but also the institution's access to federal research grants. By now many thousands of experiments have been performed using the NIH guidelines, and the earlier predictions of catastrophic consequences appear to be unfounded.

Most scientists have always viewed the risks as very small for several reasons. First, recombinant DNA, as an isolated material in a test tube, is not dangerous; only under carefully controlled conditions can it be transferred into living cells. Second, recombination (i.e., the forming of new combinations of DNA segments) occurs in nature and is not in itself dangerous. Third, bacteria do not often escape from laboratories and establish infections. Even highly evolved **pathogens** (disease-causing organisms) have been successfully contained in laboratories, and the isolated cases of laboratory infections have not led to epidemics. The strains of bacteria used for gene cloning do not survive well outside the laboratory; most of them are genetic cripples, often requiring special nutrients for growth. Fourth, the chance is small that one of these bacteria would be converted accidentally into a pathogen. Pathogenic organisms have sophisticated mechanisms for infecting their hosts. These mechanisms involve a number of different genes that have been carefully honed by millions of years of evolution. For all these reasons, prudent caution has been deemed adequate.

The controversy has now reached a second level. Should genetically engineered bacteria be deliberately released into the environment? The issue is whether the engineered bacteria will displace other bacteria from their normal ecological niches. And if so, what are the consequences, how long would we have to wait to see them, and could we reverse them?

An entirely different issue is whether we have the right and wisdom to manipulate our own genetic information and to influence the evolution of our own species, as we have done with livestock for centuries. This is not a scientific question, and, as stated, it is generally not addressed by molecular biologists. The current thrust is to correct genetic defects in particular individuals, not in the species in general. For example, individuals having genetic defects in bone marrow cells may be treated by changing the DNA in their bone marrow cells, not in their **germ cells** (**sperm** and **eggs**). Thus the individual, not his or her children, will be affected by genetic engineering. However, knowledge obtained from recombinant DNA technologies will make it possible to modify the species as a whole. It may be up to you as a citizen to decide how, or indeed whether, this type of knowledge will be applied.

PERSPECTIVE

Understanding gene cloning requires two types of knowledge: a grasp of the concepts of molecular biology and a familiarity with the necessary laboratory manipulations. Both aspects are discussed in subsequent chapters. It is important to realize that working with huge numbers of very small items requires strategies that may not initially be obvious. The molecular "scissors" used to cut DNA into small fragments makes many, many cuts. Thousands of different pieces of DNA are produced. Locating the single, desired fragment of DNA and separating it from all the other pieces is a formidable task. Unfortunately, DNA cannot be run through an editing machine like a piece of film; genes in DNA are too small to be seen by the human eye. Even with the highest powered microscopes, DNA containing thousands of genes looks like a featureless piece of string (see Figure 1-1). To find a specific gene, cloners blindly separate the DNA fragments; then they examine the different fragment types biochemically until they locate the desired ones. The separation process involves putting individual DNA fragments *randomly* into many, many bacterial cells. Thus it is necessary to describe the features of bacterial growth that make these organisms useful for isolating and identifying specific DNA fragments. This is done in Chapter 2.

Understanding how DNA itself is manipulated requires several types of information. In Chapter 3, some fundamentals of DNA structure are presented. Chapter 4 then focuses on **gene expression**, the process of converting genetic information from a DNA to a protein form, since controlling the expression of a particular gene is often important for getting the gene to work properly. The emphasis of Chapter 5 is on the reproduction of DNA, a topic that also introduces some of the protein tools used to engineer genes. Chapter 6 describes plasmids and phages, the submicroscopic vehicles that are used to carry DNA fragments into bacterial cells. Then cutting and joining DNA molecules are discussed (Chapter 7) so that the individual steps of gene cloning can be tied together into a single recipe (Chapter 8). At that point sufficient background will have been presented to permit a discussion of how gene cloning is used for determining the organization of our own genetic information (Chapter 9), as well as a few of the many interesting discoveries that have been made (Chapter 10). The final chapter focuses on three related topics relevant to everyone alive: retroviruses, AIDS, and cancer genes. By that point you will have a general understanding of one of the major strategies biologists are using to discover how living cells work.

Questions for Discussion

1. The earth and all living things on it are composed of atoms. What distinguishes living objects from nonliving ones? It cannot be growth and movement, for inanimate things such as crystals grow, and both water and air move.
2. To work with molecules, one must have ways to detect them and to measure how many are present. To experience measuring something you cannot see, first imagine that you are blindfolded and sitting at a dinner table. Then devise a way to determine how much water is in your glass without sticking your finger in.
3. Certain elements can occur as radioactive isotopes, unstable

forms in which the atoms spontaneously disintegrate. The disintegration is called radioactive decay, and it releases high energy particles and radiation characteristic of the particular type of atom. DNA and other biological molecules can be prepared in such a way that some of their atoms are radioactive. This makes it much easier to detect the biological molecule because radioactivity can be measured relatively easily. List some of the ways that radioactivity can be detected (see glossary and Chapter 8).

4. Among the elements that have radioactive forms are plutonium, hydrogen, carbon, cesium, and phosphorus. Which of these can be incorporated into biological molecules (refer to Figure 1-3)?
5. Each type of radioactive atom decays at a characteristic rate. For example, half of the atoms in the form of radioactive phosphorus that is commonly used in the laboratory decay every two weeks. Suppose that you prepared DNA containing radioactive phosphorus but had to wait two months before working with it. What fraction of the radioactivity initially in the DNA would remain?
6. We are composed of molecules. Most of our molecules are quite stable at body temperature, but within our bodies chemical reactions (conversions of one type of molecule to another) are occurring continuously. Obviously our bodies have ways to make at least some of the molecules unstable without raising the temperature to very high levels. How might this occur?
7. Discuss the ways in which gene cloning might affect your life.

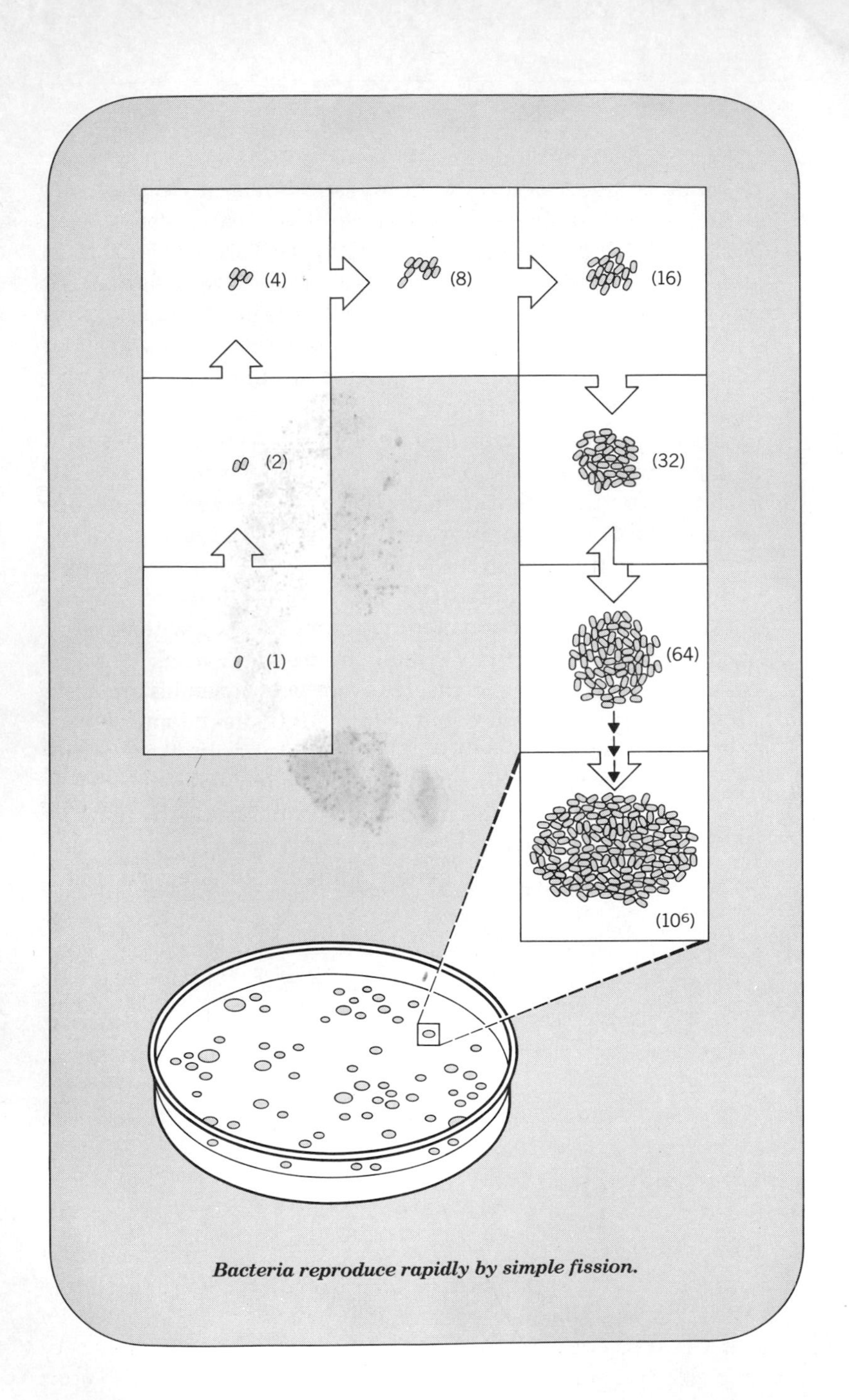

Bacteria reproduce rapidly by simple fission.

CHAPTER TWO

BACTERIAL GROWTH

Microorganisms as Tools for Gene Cloning

Overview

Bacteria are microscopic organisms that reproduce very rapidly and are easily grown in the laboratory. A single cell placed on a solid surface containing nutrients can multiply to form a colony comprised of millions of identical cells. If the colony is transferred to liquid growth medium, the cells continue to multiply; within one to two days it is possible to obtain a culture containing trillions of identical cells.

Bacteria are used in two ways for genetic engineering. First, they are used to isolate different DNA fragments. Conditions can be obtained in which a small piece of DNA will pass into the interior of a bacterial cell. The cell containing the piece of DNA can be isolated from all other cells and grown until it forms a visible colony. Second, pure cultures, in which each bacterium contains the same type of recombinant DNA, can be grown in huge quantities to produce desired protein products.

INTRODUCTION

Bacteria are one-celled organisms (Figure 2-1) that live almost everywhere: in the soil, on our skin, and in our intestines. When most people think about bacteria, diseases come to mind, diseases such as plague, botulism, anthrax, cholera, and typhoid fever. Fortunately, most bacteria are not pathogenic (disease-causing). Indeed, some of the non-

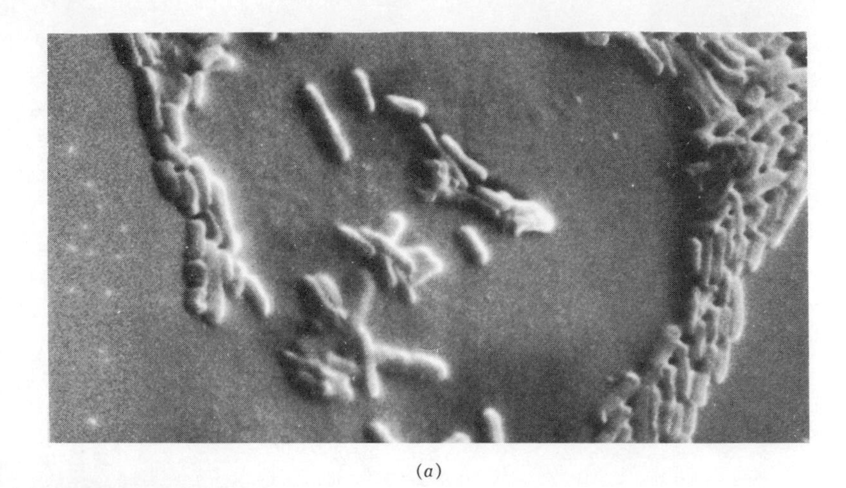

(*a*)

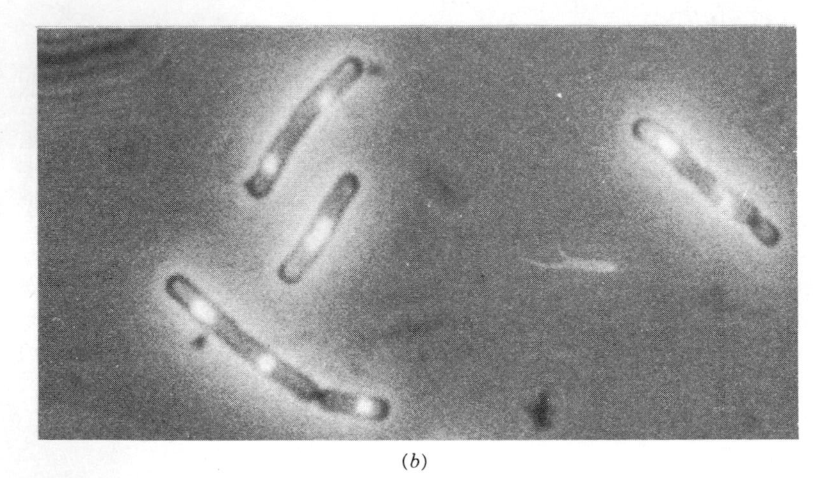

(*b*)

Figure 2-1 Photomicrographs of Bacteria. (**a**) A cluster of *E. coli* cells as they appear using scanning electron microscopy. (Magnification is 3800 times. Photomicrograph courtesy of Sandra McCormack, Rochester Institute of Technology.) (**b**) Several *E. coli* cells as they appear using light microscopy. The bright structures in the center of the cells are DNA-containing bodies called **nucleoids**, which have been stained with a dye called ethidium bromide. An electron micrograph of a nucleoid that has been removed from the cell is shown in Figure 1-1. The cells shown here are mutants that are unable to reproduce properly. As a result, the cells become elongated and appear to have more than one nucleoid per cell. (Magnification is 3800 times. Photomicrograph courtesy of Todd Steck, University of Rochester.)

pathogenic species have become the favorite experimental subjects of molecular biologists, and the study of these microbes has led to gene cloning technologies. The two sections that follow outline the growth properties of bacteria that make these microorganisms so useful. The final section of the chapter briefly describes some of the life processes that occur inside bacteria to help clarify the role of DNA.

CHARACTERISTICS OF BACTERIAL GROWTH

About 45 years ago, molecular biologists became interested in bacteria because these organisms are very tiny (a high power microscope is needed to see individual bacteria) and because bacteria seem to lead such uncomplicated lives. They simply grow and then divide in half to produce two new cells. Each of the new, daughter cells then expands until it, too, divides to form two more cells. Thus bacteria reproduce by simple **fission**. This simplicity produced the hope that everything about bacterial life could be understood in molecular terms, that the essence of life itself could be understood.

Two properties of bacteria are particularly important. First, each bacterium is only a single cell, lacking limbs, organs, and complicated developmental stages. Consequently, it is relatively easy to obtain large numbers of *identical* cells to study cell chemistry. Second, bacteria grow and multiply rapidly. Experiments that would take years with other organisms can be done in a single day with bacteria. For example, in the laboratory many kinds of bacteria divide every 40 minutes. Thus, during an 8-hour day, a batch of reproducing bacteria will go through 12 generations, greatly facilitating the study of how traits are passed from generation to generation.

The two properties mentioned above, small size and rapid growth, make it easy to cultivate large numbers of bacteria. The recipe for growing bacteria is simple: place a few bacteria in lukewarm broth, and let the bacteria do the rest. They grow and divide, producing what is called a **bacterial culture**. Within 24 hours even a small flask of broth may contain billions of bacteria.Thus it is easy to get astronomical numbers of bacteria.

Although the growth properties of many bacteria types make them suitable tools for gene cloning, a species called *Escherichia coli* (***E. coli*** for short) is used most extensively. *E. coli* is not particularly distinctive; it is a small rod-shaped organism (Figure 2-1) that is normally a harmless inhabitant of the human digestive tract. Like most other bacteria, its shape is maintained by a rigid coating called a **cell wall**. Inside this wall the chemical reactions of life take place. *E. coli* is special because 40 years ago a group of molecular biologists focused their research on it. As details about the chemistry of *E. coli*'s life were learned, it became easier to conduct more sophisticated experiments on this bacterium than to start over with a new organism. As a result of this intense effort, *E. coli* has become the best understood organism on earth. Likewise, the best known cloning vehicles are infectious agents that use *E. coli* as a host, for they too have been intensively studied for many years.

Since our host for cloning, *E. coli*, is so small, special methods are required to use it. One cannot simply look through a microscope to see whether a particular cell has taken up a specific piece of DNA; the cells have few distinctive characteristics when viewed through an ordinary microscope. Only under special conditions can the DNA be seen inside the cell, and then it looks like a featureless blob (Figure 2-1*b*). Nor can one dissect a bacterium, as many of us dissected frogs in general biology class. Instead, indirect observations must substitute for what cannot be seen. For example, a liquid bacterial culture will appear cloudy if it contains more than 10 million cells per cubic centimeter (a quarter-teaspoon). As the bacteria multiply, the cloudiness increases. Thus, by measuring how fast the cloudiness increases, it is possible to determine the rate at which the bacteria are reproducing.

BACTERIAL COLONIES

A single bacterial cell growing on a solid surface will multiply to form a cluster of cells called a **colony**. Since this is one of the more important concepts in gene cloning, a procedure is presented below for obtaining bacterial colonies, a procedure that could be carried out in almost any kitchen. Biologists use slightly more refined equipment to obtain colonies, but the principles are the same. First **dissolve** in boiling water some **agar**, which is a gelatinlike substance. Add sugar, minerals, and perhaps a rich source of nutrients such as beef extract. Pour the resulting solution into a **sterile**, covered dish and set aside, allowing the agar to cool and solidify. Next place a teaspoon of soil in a tablespoon of water. Stir briefly. Soil is a good place to obtain microorganisms, since many types of bacteria live there. Bacteria such as *E. coli* can be obtained from sewage.

Now find a thin piece of wire and bend the end to form a loop about a millimeter in diameter. Heat the loop with a flame to sterilize it. Dip the loop directly in the mixture of soil and water and lift it out. The loop will trap a tiny drop of water containing bacteria. Place the drop of water from the loop onto the surface of the solid agar, and smear the drop over the surface of the agar with the wire loop. Replace the lid on the dish, and allow the bacteria to grow and multiply at room temperature. Several days later hundreds of small bumps will be seen on the surface of the agar. These bumps, which often look like glistening blisters about a millimeter across (see Figure 2-2 and frontispiece for Chapter 2), are bacterial colonies.

When the drop of water was smeared over the agar, a small number

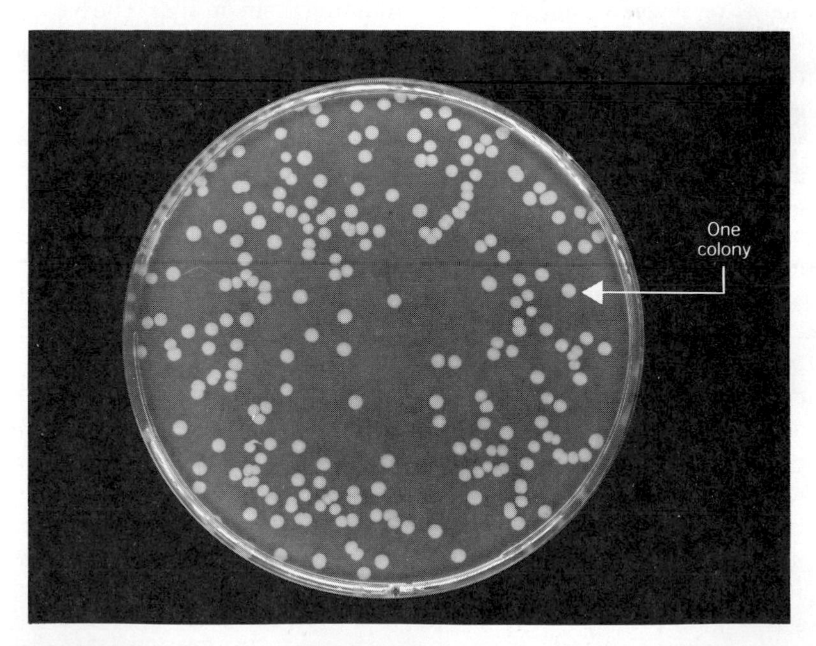

Figure 2-2 Bacterial Colonies Growing on Agar. A dilute suspension of *E. coli* cells was spread on solid agar in a petri dish and was incubated at 37° C. After 24 hours, colonies, each about 2 millimeters in diameter, became visible on the agar. The arrow points to one such colony.

of bacteria from the soil were scattered to widely separated spots on the agar. Each cell divided many times, and since the new cells could not move away from each other, they piled up. Within a day or so, the bacterial colony became visible. With this simple procedure it is easy to separate and culture the individual bacterial cells in the original soil sample. *The millions of bacteria in a colony all arose from a single bacterial cell.* Thus all cells in a colony are identical; they are members of a clone.

The ability to obtain individual colonies is important because gene cloners cannot visually distinguish one gene from another. We cannot simply use forceps to pick particular genes out of a pile of DNA fragments. Instead, we use bacteria. Bacteria can incorporate small pieces of DNA if the DNA is linked to a cloning vehicle, a small, **infectious** DNA molecule. Thus gene cloners isolate DNA fragments by first transferring them into bacteria and then separating the bacteria from each other by spreading the broth containing them on agar. The bacteria grow into visible colonies, and the cloned DNA fragments multiply millions of times. The colonies are then tested for the presence of particular DNA fragments as described in later chapters.

Finding the colony that contains a particular gene is not easy; the gene cloner faces a problem of numbers. First, the gene being sought may be rare; it is not uncommon for the DNA fragment containing a particular gene to represent less than one out of 100,000 DNA fragments. Second, transferring DNA into bacterial cells is an inefficient process. In some cases fewer than one in 10,000 cloning vehicles will take up residency in a cell, and only a fraction of these will be attached to one of the fragments being sought. Thus the chance that any particular bacterial cell will contain a specific fragment may be less than one in a billion. The cloner's main task is to find that rare cell.

Whenever biologists work with an organism, they try to obtain a pure culture, one that contains only the type of organism being studied. Interpreting experimental results or producing a pure product is very difficult if other types of organism contaminate the culture. Our ability to grow single bacterial colonies makes it easy to obtain pure cultures. The principle of pure cultures can be illustrated by describing one way to clone the insulin gene using bacteria. First, human DNA is obtained (a method is presented in Chapter 8). The DNA is cut into millions of discrete fragments and inserted into cloning vehicles, producing many different types of recombinant DNA molecule. The collection of recombinant DNA molecules is next mixed with a huge number of *E. coli* cells. Some of the recombinant DNA molecules get inside bacterial cells. The cells are then spread out on the surfaces of agar plates to separate one cell from another. Since the cloning vehicle contains a gene for antibiotic resistance, addition of the appropriate antibiotic to the agar allows growth only of cells that contain the cloning vehicle. After colonies have arisen, each is biochemically tested by strategies described in later chapters until a colony containing insulin genes is found. Then that particular colony is carefully touched with a piece of sterile wire in such a way that some of the bacterial cells in the colony stick to the wire. The cells on the wire are then transferred into a flask of sterile broth by simply dipping the end of the wire into the broth. Some of the cells fall off the wire and begin to grow and divide in the broth. By the next day the flask will be full of bacteria, each of which has arisen from the single bacterial cell carrying the insulin gene. The flask contains what microbiologists call a pure culture of bacteria, one having only a single type of organism; all the cells in the flask are members of a clone. This culture can be maintained indefinitely if care is taken to keep other bacteria out of the culture. To grow large amounts of the bacteria, one simply transfers a drop from the pure culture into a vat of sterile broth. Within a few days the vat will contain trillions of bacteria, each containing the insulin gene.

BIOCHEMICAL ASPECTS OF BACTERIAL GROWTH

Bacterial cells contain elaborate mechanisms that allow them to reproduce. A knowledge of how this machinery works is central to understanding genetic engineering, for most of the gene cloning tools, including the genes themselves, are components of this machinery. We can begin to dissect the machinery by first focusing on the relationships among cellular chemical reactions, **enzymes**, and DNA.

Although people have many definitions of life, one experimentally useful view is that life is a collection of chemical reactions organized to reproduce itself. In these reactions molecules are broken down into simpler molecules, built up into more complicated molecules, or simply rearranged to create slightly different molecules. Over the years chemists have been able to describe a large number of the rules that govern these reactions. In some ways a series of chemical reactions is similar to tailoring a shirt: the cloth is cut into specific shapes, the front and back are joined together, the sleeves and collar are stitched, and then they are attached to the body. Finally buttons are added. Each step is analogous to one of the chemical reactions in the series; through a series of steps the original piece of cloth is converted into a new form.

Although one person working alone can make a shirt, the analogy to cell chemistry is clearer if we think of each task as being the assignment of a particular workman—a cutter, a stitcher, and so on. If shirts are to be produced efficiently, the tasks must be performed in a particular order. So it is with the work of enzymes, the specialized protein molecules that control the chemistry of cells. In *E. coli* there are more than a thousand DIFFERENT types of enzyme that control production of the cellular components. The key idea is that the chemicals in *E. coli*, as well as those in every cell in our bodies, react in an orderly fashion; it is the enzymes that provide that order. Some enzymes are very valuable engineering tools, particularly those that cut DNA and those that join DNA molecules together.

Most enzymes are proteins (a few RNA molecules have been discovered that act as enzymes; they are called **ribozymes**, and they are briefly discussed in Chapter 10). Proteins are long, linear molecules that are much like beaded necklaces folded in a specific way (Figure 2-3). Some proteins contain hundreds or even thousands of beads. Each bead is called an **amino acid**. There are 20 different kinds of amino acid, so there are an astronomical number of ways in which a chain can be put together. Consequently, it is easy to show that there could exist many, many different kinds of protein. The chemical properties of the amino acids in a protein determine precisely how the protein folds and thus how it acts.

Enzymes are catalysts. They speed up chemical reactions without themselves being consumed in the reaction. Many enzymes do this by binding to specific molecules in such a way that bonds between certain atoms in the molecules can be formed or broken more readily than in the absence of the enzyme. In many cases enzymes contain pockets or clefts on their surfaces into which fit the molecules involved in the reaction.

It is important to realize that the properties of a particular enzyme, that is, how it controls a particular chemical reaction, are determined by the *order* of the amino acids in the chain. Consequently, knowing how the cell specifies the precise order of amino acids in a protein is

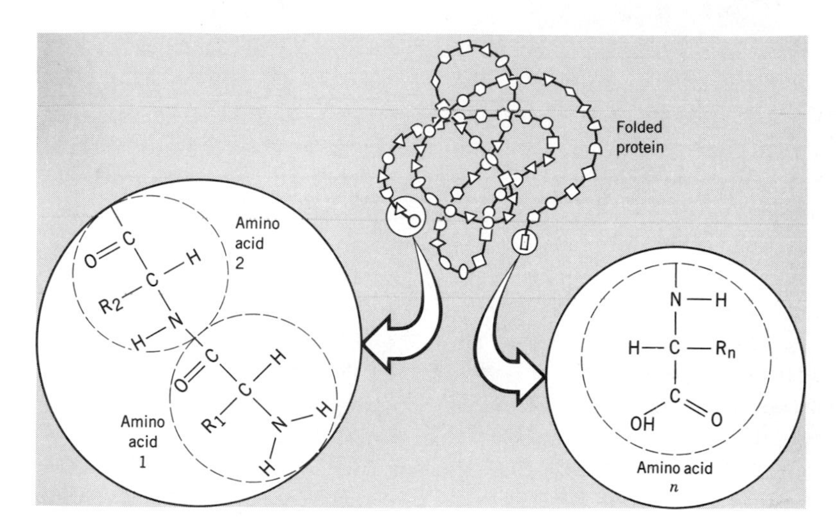

Figure 2-3 Protein Structure. Proteins are long chainlike molecules composed of many amino acids. There are 20 types of amino acid, which are distinguished by the arrangement of atoms in the group labeled **R** in each amino acid. The precise folding of the protein chain is determined by the arrangement of R groups. Each protein has chemically distinct ends. The amino acid labeled **1** is called the amino terminal amino acid. Note that its nitrogen has one more hydrogen than that of amino acid **2**, and its nitrogen is bonded to only one carbon atom. Amino acid **n** is the carboxy terminal amino acid. At this end of the protein the terminal carbon is bonded to two oxygen atoms.

central to understanding how enzymes are made. Such knowledge is the key to understanding heredity, for the characteristics of enzymes and other proteins make a cell what it is. The information determining the order of amino acids in every protein is stored in DNA. How the information in DNA is converted into protein (Figure 2-4) will be discussed in more detail in Chapter 4.

PERSPECTIVE

Throughout our history bacteria have plagued us by causing diseases. We have learned to live with these tiny organisms by washing our hands, treating our drinking water, immunizing our bodies, and keeping bacteria-infected fleas from biting us. Occasionally our efforts fail,

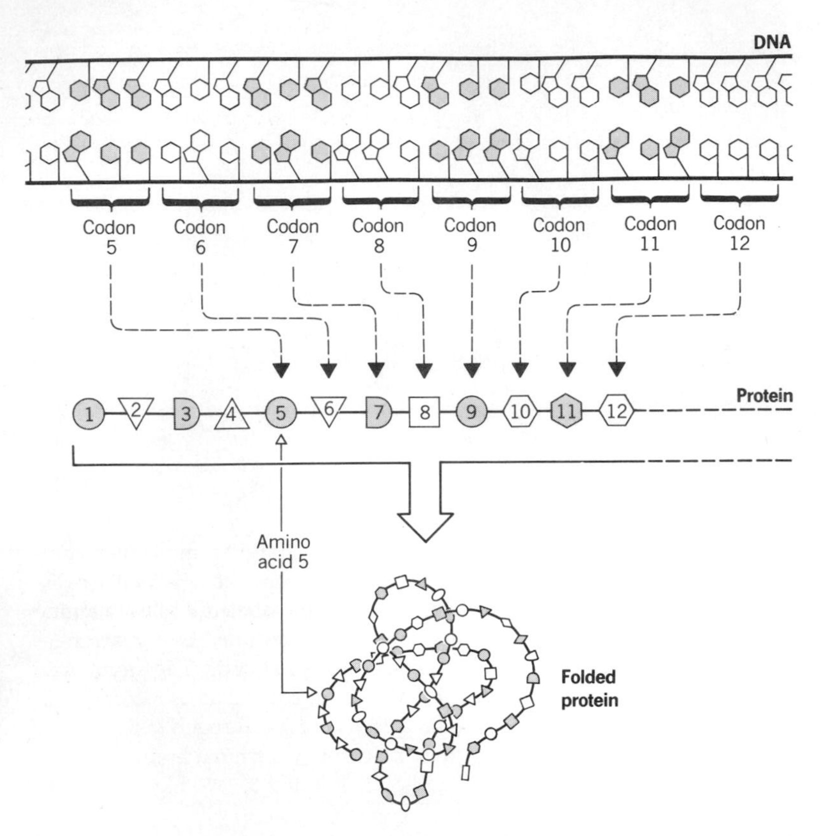

Figure 2-4 Relationship of Information in DNA to Protein Structure. Information in DNA is arranged in a series of 3-letter words called ***codons***. Each codon specifies a particular amino acid. For example, codon 5 has information for the fifth amino acid in the protein. The overall shape and activity of a protein depend on the precise order of amino acids. The information for this order is stored in the DNA. The two DNA strands are wound around each other (see Figure 3-1); for clarity, winding of the DNA strands is not shown here. Also omitted for simplicity are messenger RNA and other aspects of how information in DNA is converted into protein (see Chapter 4).

and we have to kill bacteria with antibiotics. We have also learned to use bacteria. For example, bacteria help us make products such as yogurt. Recently, our relationship with bacteria has entered a new phase: we are now using these organisms to help understand and manipulate the chemistry of life.

It turns out that many of the biochemical properties of bacteria are common to all life forms, including man; consequently, it is possible to describe bacteria in the context of gene cloning and also to introduce the concepts of molecular biology. There are, however, fundamental differences between bacterial cells and our own cells. For example, in all higher organisms, from yeast to humans, DNA is contained in a membrane-bound body called a nucleus. Bacterial DNA is not held in a nucleus. Other differences will be pointed out in subsequent chapters, but the focus will be on features shared by all organisms.

Questions for Discussion

1. Bacteria are about 1 to 2 **micrometers** (μm) long. Compare their size to other small objects with which you are familiar.
2. In broth containing many nutrients some bacteria grow rapidly, dividing every 20 minutes. At this rate, how many cells would there be in a culture after 10 hours of growth if the culture started with only one cell?
3. A bacterial colony contains roughly one million to ten million cells. Describe how you would determine the number of live cells in a colony growing on an agar plate using only a spatula to scrape off the colony, test tubes, liquid growth media, pipettes to measure liquid volumes, agar plates, and an incubator?
4. How many ways can 20 different types of amino acid be arranged in a peptide (a chain of amino acids) 10 amino acids long?
5. The chemical structure of DNA is changed (damaged) by radiation (X-rays, ultraviolet light, etc.). Sometimes the radiation can even break DNA. How do you think DNA breaks would affect bacterial growth? How do you think DNA breaks arising from sunlight might affect your skin?

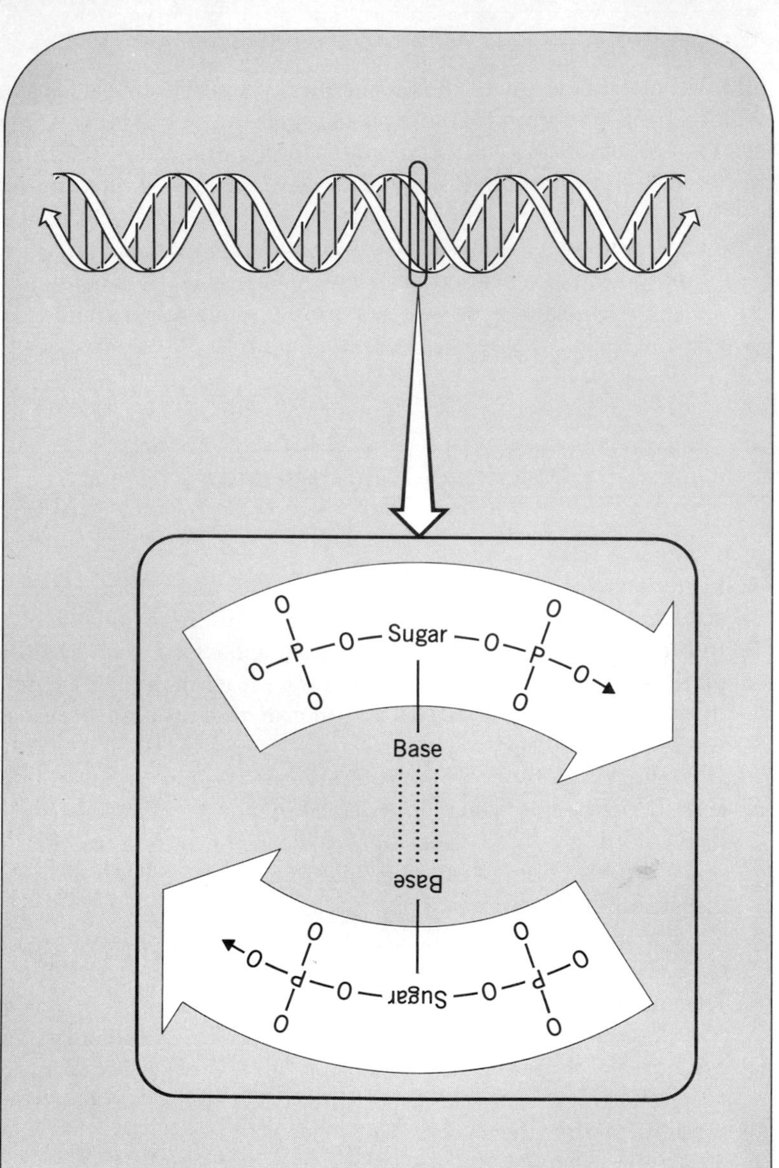

In DNA the strands run in opposite directions and are held together in part by interstrand interactions between the bases of the nucleotides.

C H A P T E R   T H R E E

STRUCTURE OF DNA

Two Long, Interwound Chains

Overview

DNA is a long, threadlike molecule composed of subunits, and in a sense it resembles a two-stranded beaded necklace. DNA molecules can be very long, sometimes containing more than a hundred million subunits called nucleotides. There are four different types of nucleotide (abbreviated A, T, G, and C), and it is through the specific order of the nucleotides that genetic information is stored in DNA. The nucleotides in the two strands pair with each other so that an **A** in one strand is always opposite a **T** in the other, and a **G** is always opposite a **C**. This rule allows the two strands to act as templates for the formation of new strands. The two ends of a DNA molecule differ in chemical reactivity. Unlike a rope, DNA has distinct left and right ends, and the two strands run in opposite directions.

Genes are discrete stretches of nucleotides that contain information specifying the sequence of amino acids in proteins. It takes three nucleotides to specify a particular amino acid; that is, specific nucleotide triplets or codons correspond to specific amino acids. Specific combinations of nucleotides also signal the beginning and end of a gene. In terms of information content, DNA is a very stable molecule, and the information is faithfully reproduced and passed from one generation to the next. But DNA should not be considered to be a static molecule in terms of its three-dimensional structure, for it interacts with many cellular proteins that can dramatically alter its shape.

INTRODUCTION

To understand how DNA controls the activities of our cells and thus the activities of each of us, it is necessary to briefly discuss the structure of DNA. One level of discussion focuses on the chemical structure of the subunits, the nucleotides. The structure of the nucleotides holds the key to understanding how DNA is able to reproduce and how information contained in DNA is utilized. Another level of structure is the order of the nucleotides. It is the order that determines the information content, and it is the order that determines in part where specific proteins will bind and cause DNA to carry out its many activities. How the order of the nucleotides is determined experimentally will be described in Chapter 9. A third aspect of structure deals with DNA as a whole, as a long, flexible molecule. Three-dimensional structure is important when considering how the cell gains access to information in specific regions of DNA because twisting and wrapping of DNA can greatly alter the ability of proteins to bind to it.

CHEMICAL STRUCTURE OF DNA

In earlier chapters the organization of information in DNA was described by drawing parallels between it and motion picture film. But to describe the chemistry of DNA, other analogies must be developed. One can think of DNA as a long, thin string composed of two strands wound around each other much like strands in a rope (Figure 3-1*a*). But closer inspection, using biochemical rather than microscopic methods, reveals that each strand is composed of tiny subunits. Thus, the DNA strands are more accurately thought of as two interwound strings of beads, with each string containing millions of minute subunits linked together (Figure 3-1*b*). Chemists call the subunits **nucleotides**, and they have found that each nucleotide is composed of three parts (Figure 3-2): a flattened ring structure called a **base**, a **sugar** ring called deoxyribose, and a **phosphate**. Alternating sugars and phosphates form the backbone of the DNA. The bases are located between the backbones of the DNA strands, and they lie perpendicular to the long axis of the strands (Figures 3-1*c* and 3-3). As the backbones of the two strands wind around each other, they form a double helix (Figures 3-1*c* and 3-3), leading to the popular expression for DNA.

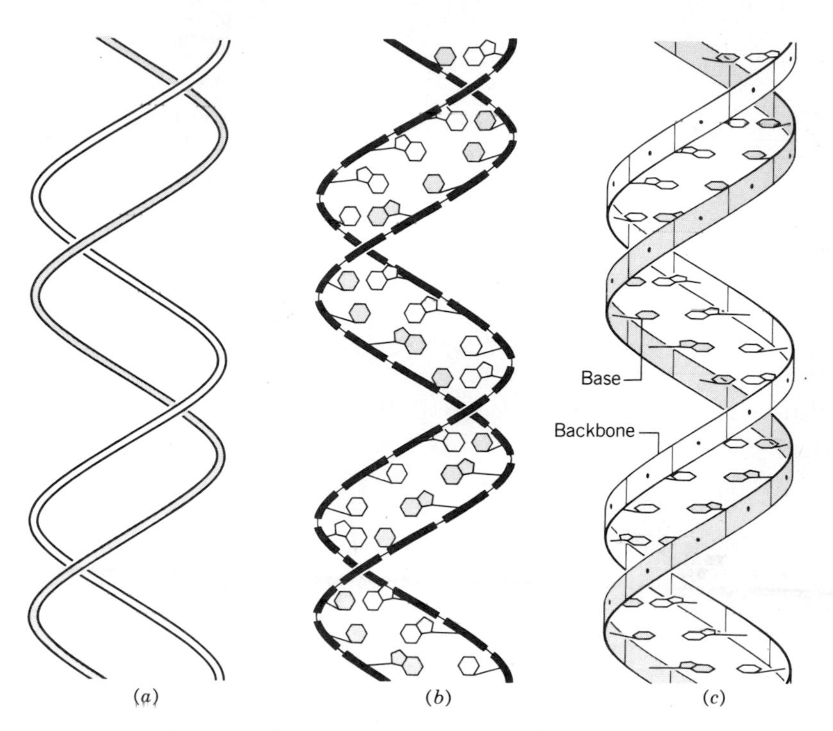

Figure 3-1 Schematic Representations of DNA. (a) Two interwound strands. **(b)** Two interwound beaded chains. **(c)** Double helix of two interwound strands with bases on the inside and backbone on the outside.

The bases tend to stack one on top of another, much like steps in a spiral staircase. The bases come in four varieties, popularly abbreviated A, T, G, and C. The letters stand for **adenine, thymine, guanine,** and **cytosine,** the chemical names of the bases. Since each nucleotide contains only one base, the nucleotides can also be identified by the same four letters. These four nucleotides are precisely ordered in DNA, and it is through this arrangement of nucleotides that cells store information. The principle is similar to the Morse Code, where information is transmitted in combinations of two symbols, dots and dashes.

Extensive examination of DNA has led to the identification of three rules that govern DNA structure. First, a single DNA strand does not have branches. Consequently, the information is stored in a simple

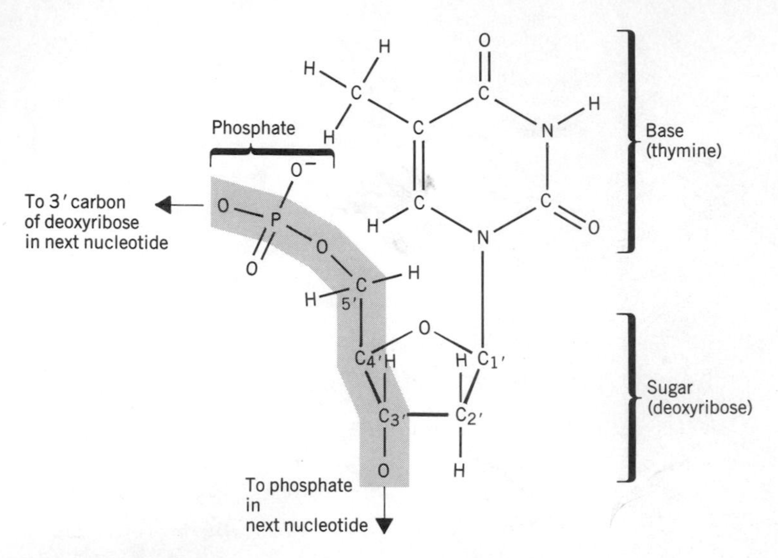

Figure 3-2 Structure of a Deoxyribonucleotide. A nucleotide is composed of three parts, the base (in this case it is thymine), the sugar deoxyribose (see also Figure 1-4), and a phosphate. One nucleotide is joined to the next by bonds between deoxyribose and phosphate. The carbon atoms of the deoxyribose are numbered 1′ through 5′. The 3′ carbon of the sugar attaches to a phosphate from the nucleotide immediately to its right in DNA. The phosphate bound to the 5′ carbon of the sugar attaches to the nucleotide at its left by binding to the 3′ carbon of the sugar in that nucleotide. The shaded region represents the atoms that are a part of the backbone of DNA.

line. Second, the ends of a DNA strand are chemically different. Thus a strand of DNA has directionality. As shown in Figure 3-4, an end is named according to the sugar carbon at that end, and by convention the 5′ end is generally drawn as the left end. Third, when two DNA strands come together and form a double helix, bases must fit together in a precise way. Whenever an A occurs in one strand, a T must occur opposite it in the other strand. Likewise, G always aligns opposite C. Only when the bases are properly paired will the two DNA strands fit together. This third rule is called **complementary base pairing**. It is important to note that the two strands of DNA are complementary, NOT identical; identical nucleotides do not form base pairs. One can imagine that the bases opposite each other fit together like electrical plugs and sockets (Figure 3-5); only the correct pairs coincide. As a re-

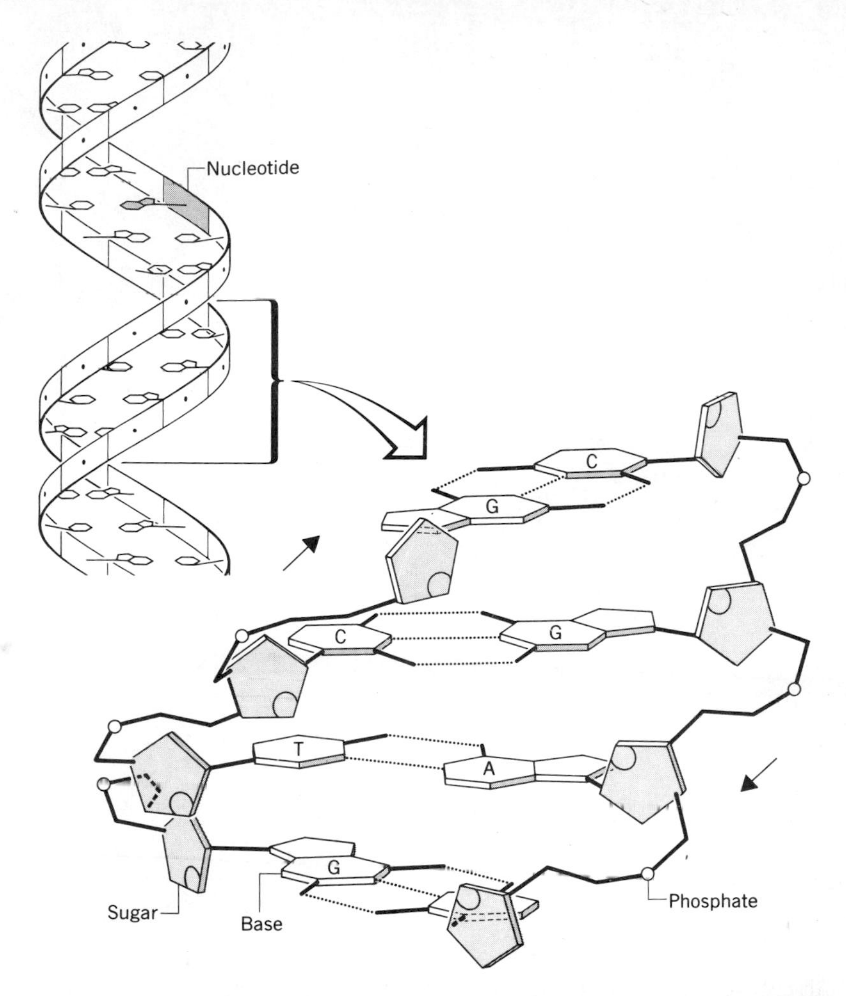

Figure 3-3 Schematic Diagram of a Short Section of a DNA Double Helix. Each DNA strand is composed of chemical structures of three types : bases, sugars, and phosphates. A nucleotide is a unit composed of one base, one sugar, and one phosphate. The sugars and phosphates connect to form the backbone of each strand, and a base attaches to each sugar. The four different bases are represented by the letters A, T, G, and C. The bases of one strand point inward and toward those of the other. Attractive forces called **hydrogen bonds** (represented by broken lines) exist between the bases of opposite strands and contribute to holding the two strands together. The two strands run in opposite directions; notice how the sugars in one strand seem to point upward while those in the other seem to point downward. See Figure 3-4 for additional description of directionality.

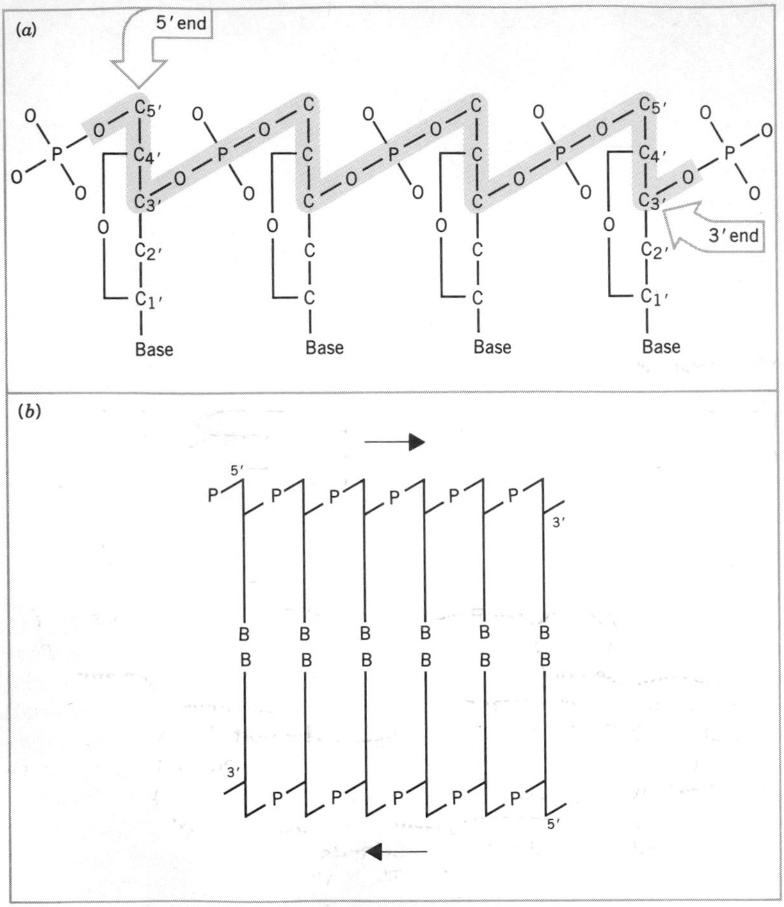

Figure 3-4 **Nucleic Acid Directionality.** (**a**) The ends of nucleic acids are distinct and have different chemical properties. For illustrative purposes, the sugar portion of the nucleic acid has been stretched out and only its carbon (C) and oxygen (O) atoms are shown. The sugars are joined by phosphates, and the backbone of the chain is illustrated by the shaded region. The five carbon atoms in each sugar group are assigned numbers 1′ through 5′. The ends of the nucleic acid are designated by the symbols 5′ and 3′. At the 5′ end of the chain a terminal phosphate (PO_4) is joined to sugar carbon number 5′, and at the 3′ end a terminal phosphate is connected to sugar carbon number 3′. A nucleic acid can end with either a phosphate group or a hydroxyl (OH) group; enzymes that act on the ends of a nucleic acid often prefer a particular type of end. (**b**) The two DNA strands have opposite polarity. Here, deoxyribose is reduced to a line (see Figure 1-4e), phosphate to a **P**, and base to a **B**. By convention, the upper strand has its 5′ end on the left.

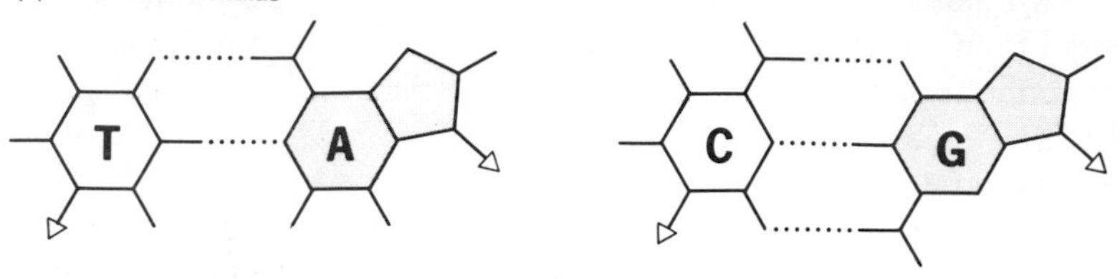

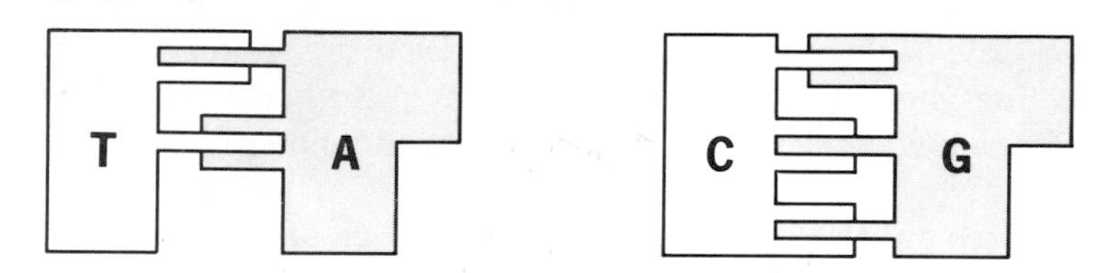

Figure 3-5 Complementary Base Pairing. (**a**) Structural formulas for thymine:adenine (TA) and cytosine:guanine (CG base pairs). The bases are flat structures composed of hydrogen, carbon, nitrogen, and oxygen atoms. The solid lines represent chemical bonds between these atoms. Arrows indicate points at which the bases attach to sugars. The dotted lines are hydrogen bonds, weak attractive forces between hydrogen and either nitrogen or oxygen. Notice that there are two hydrogen bonds between adenine and thymine, and three between guanine and cytosine. The difference in hydrogen bonding is the structural explanation for complementary base pairing. (**b**) A prongs-and-sockets analogy for base pairing. The hydrogen atoms in each hydrogen bond are represented as prongs, and the oxygen or nitrogen atoms are depicted as sockets. The attractive forces are weak; consequently, perfect fits are required for base pairing to occur.

sult of the millions of tiny bases fitting together and stacking on top of each other, the two strands of DNA tend to stick tightly together. Thus the DNA double helix is a stable structure; temperatures near that of boiling water are required to separate the strands.

ORGANIZATION OF INFORMATION IN DNA

As pointed out above, the subunits of the DNA strands, the nucleotides, are the chemical basis for storage of information in DNA. Returning to the film analogy introduced in Chapter 1 (Figure 1-2), the units we have now defined as nucleotide pairs, or base pairs, correspond to the frames in a motion picture film. That is, the "genetic letters" mentioned earlier represent the chemical base name abbreviations A, T, G, and C. Thus, information is stored in DNA by the specific sequence of a 4-letter code (A, T, G, C), which reads in a line along the DNA like frames in a film. In physical terms, a gene is a stretch of DNA ranging from a few hundred to a few thousand nucleotide pairs; it corresponds to a scene in the motion picture film.

We can begin to define a gene more precisely by considering features important for converting information from DNA into protein. First, a gene has a beginning and an end; there are short stretches of nucleotides that signal where the gene starts and other short segments that indicate where it stops. Second, the information in a gene is arranged as words rather than as individual letters. This is because proteins, which are also linear chains, are composed of subunits (amino acids) of 20 kinds, which are chemically different from the 4 subunits (nucleotides) used to store information in DNA. A quick calculation predicts how many nucleotide letters would be necessary to code for each amino acid, that is, how many letters in DNA correspond to each letter in a protein. Obviously one nucleotide in DNA cannot correspond to one amino acid in protein because there are 20 different amino acids and only 4 different nucleotides. Likewise, the nucleotides cannot be read in pairs because 4 different nucleotides taken two at a time can produce only 16 possible pairs, 4 short of the minimum number. The code can be read in threes: 4 nucleotides taken three at a time give 64 possible triplets. This is more than enough to specify the 20 amino acids as well as the necessary punctuation, such as start and stop signals. Many experiments have confirmed the prediction that the genetic code is read as triplets of nucleotides.

Triplets in DNA that correspond to amino acids are called **codons**. When the code is read by the machinery of the cell, it is first converted into an **RNA (ribonucleic acid)** form, which is similar to DNA but with the base **uracil** (U) substituted for the thymine (T) in DNA. The genetic code, in its RNA form, is shown in Figure 3-6. The conversion process is called **transcription**, and the RNA is called **messenger RNA**. The information in messenger RNA (mRNA) is next converted into amino acid sequences in protein by a process called **translation**. Both transcription and translation are described in Chapter 4.

There is no punctuation between the codons. Thus it is important

First base	Second base: U	Second base: C	Second base: A	Second base: G
U	UUU, UUC: Phe UUA, UUG: Leu	UCU, UCC, UCA, UCG: Ser	UAU, UAC: Tyr UAA, UAG: TERM	UGU, UGC: Cys UGA: TERM UGG: Trp
C	CUU, CUC, CUA, CUG: Leu	CCU, CCC, CCA, CCG: Pro	CAU, CAC: His CAA, CAG: Gln	CGU, CGC, CGA, CGG: Arg
A	AUU, AUC, AUA: Ile AUG: Met	ACU, ACC, ACA, ACG: Thr	AAU, AAC: Asn AAA, AAG: Lys	AGU, AGC: Ser AGA, AGG: Arg
G	GUU, GUC, GUA, GUG: Val	GCU, GCC, GCA, GCG: Ala	GAU, GAC: Asp GAA, GAG: Glu	GGU, GGC, GGA, GGG: Gly

Figure 3-6 The Genetic Code. Four nucleotides in RNA, taken three at a time, can form the 64 combinations shown: 61 correspond to amino acids in protein; the other 3 are stop (termination) codons. The amino acids are listed by the 3-letter abbreviations given in the **amino acid** entry in the glossary. Notice that all amino acids except tryptophan and methionine have more than one codon. Where there are multiple codons for amino acids, the base in the third position seems to have the least meaning. The bases in each codon are written with the 5′ end of the RNA on the left and the 3′ end on the right (see Figure 3-4 for description of 5′ and 3′ ends). In its DNA form, the code would contain T's instead of U's. Adapted from P. B. Weisz and R. N. Keogh, *The Science of Biology*, McGraw Hill, New York.

that the **reading frame** of the nucleotide code be established correctly —the start signal must be in the right place. This principle can be illustrated by the following sentence read as 3-letter words:

JOE SAW YOU WIN THE BET

If you read in sets of three but start out of register at the letter O instead of J, you would end up with a meaningless sentence:

OES AWY OUW INT HEB ET

Consequently, whenever an engineer inserts genes into a new DNA molecule, the correct start signal must be present to establish the right reading frame, both for the amino acids being joined and for the stop signal.

DNA AS A LONG FIBER

Chromosomal DNA molecules are often more than a thousand times longer than the cell in which they reside. Thus DNA must be folded and compacted. One level of organization is the folding of DNA into large loops, each containing about 100,000 nucleotide pairs. Another level, found in higher organisms, is the wrapping of 200 nucleotide-pair stretches of DNA around specific proteins called **histones**. This wrapping generates a series of ball-like structures along the chromosome. A similar wrapping of DNA is thought to occur in bacteria, but the ball-like structures are not readily detected in these cells.

Many activities of DNA involve pulling the strands apart, at least temporarily. Since the strands are wound around each other, strand separation requires the ends to rotate. But many DNA molecules are circles, lacking ends for rotation. To overcome this problem, cells contain enzymes called **topoisomerases**. These enzymes can cut DNA, allow intact strands of DNA to pass between the cut ends, and then seal the ends back together. This type of activity can introduce and remove twists in DNA, tie and untie DNA knots, and even link or unlink circular DNA molecules. Thus DNA is a flexible, dynamic molecule that responds in a variety of ways to the action of proteins inside the cell.

PERSPECTIVE

In 1953 James Watson and Francis Crick made what is considered to be among the most important scientific findings of modern biology. They proposed a structure for DNA that has guided the thinking of biologists to the point that the chemistry of heredity can be clearly explained and manipulated. Determining the information content, the nucleotide sequence, of DNA molecules is routine, and the sequence for many small DNA molecules is completely known. It is taking longer to determine the sequence of the larger molecules, such as bacterial and human chromosomal DNAs, but the progress is rapid. Already more than 20% (800,000 nucleotides) of the *Escherichia coli* chromosomal DNA sequence is publicly available through a computer repository called Genbank.

DNA does not exist inside the cell as a naked molecule: it is constantly interacting with proteins, many of which have a profound effect on how the DNA twists and loops. Although these structural changes do not influence the information content—that is, the sequence of nucleotides—they do affect how other proteins interact with DNA. Thus, they affect the use of genetic information by the cell, a topic that will be discussed in Chapter 4.

Questions for Discussion

1. What is a helix and why is the DNA molecule called a double helix?
2. Genetic information is encoded in the sequence of nucleotides in DNA, and within a decade or so biologists may know the nucleotide sequence for the human genome. Will this information tell us everything we need to know to cure human genetic diseases?
3. Hydrogen bonds are weak chemical interactions that are largely responsible for complementary base pairing. If there are an equal number of AT and GC base pairs in a DNA molecule 4 million base pairs long, how many base-pairing hydrogen bonds would there be in the molecule? (Hint: see Figure 3-5.)
4. Atoms sometimes carry electrical charges. Two atoms of different charge (plus and minus) attract, and atoms with the same charge (both plus or both minus) repel each other. The phosphates in DNA are negatively charged. What effect does this have on the ability of double-stranded DNA molecules to maintain a double-stranded structure? How might the binding of positively charged proteins to DNA affect the ability of DNA to be double-stranded?
5. The genetic code is read as triplets, and often several triplets encode the same amino acid (see Figure 3-6). Three triplets also sig-

nal the end of a protein (listed as TERM in Figure 3-6). Show how a single base change can convert codons for tyrosine, cysteine, tryptophan, leucine, and lysine into termination codons.

6. Many of the activities of nucleic acids depend on their ability to recognize each other. Recognition is based on the principle of complementary base pairing (see Figure 3-5). Thus base pairing must be a property common to all living organisms. Base-pairing forces are not very strong, and temperatures near boiling water are sufficient to break the base pairs. This causes double-stranded nucleic acids to become single-stranded. With these points in mind, explain how some organisms can grow in thermal pools at temperatures near boiling? (Hint: question 4 may be helpful.)

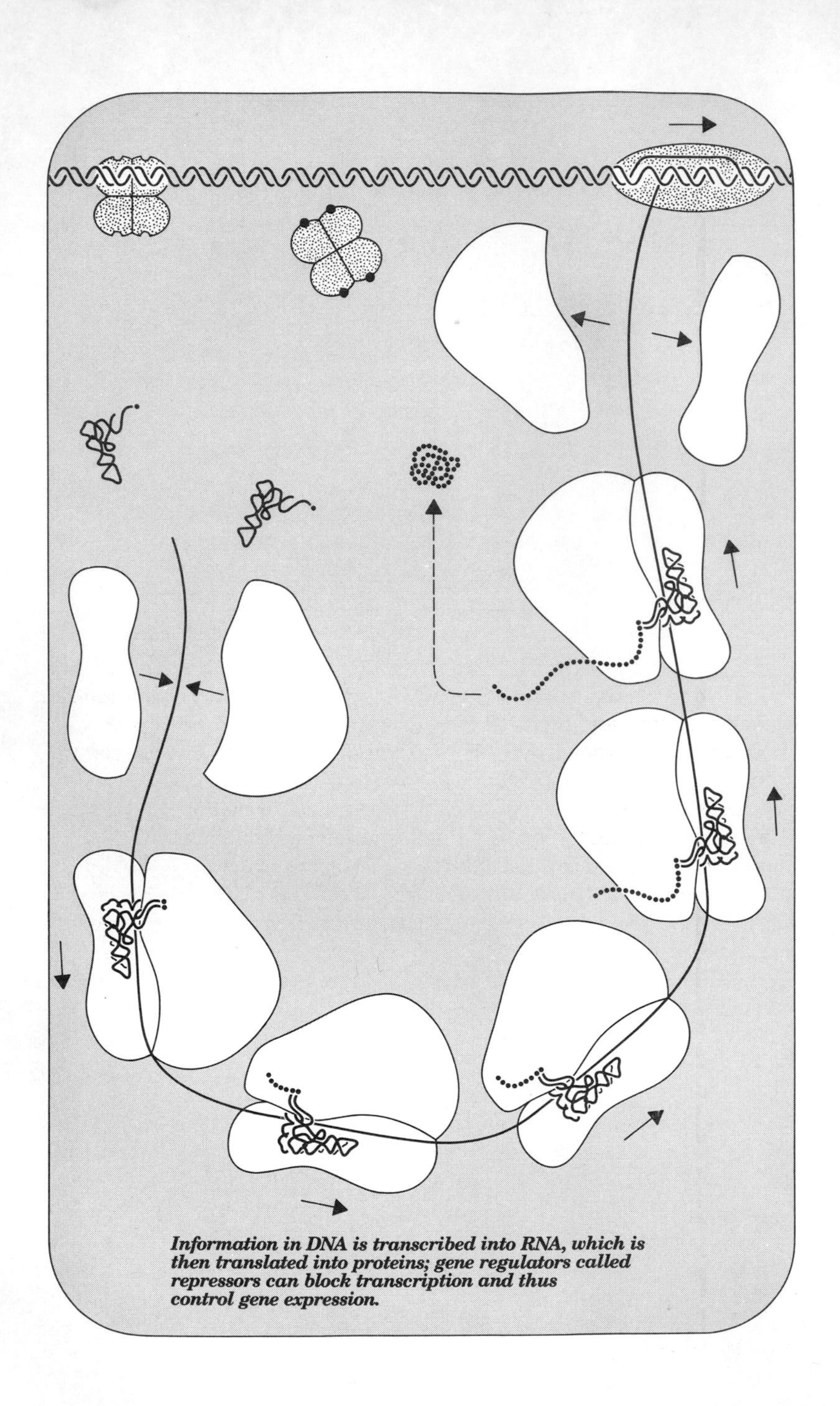

Information in DNA is transcribed into RNA, which is then translated into proteins; gene regulators called repressors can block transcription and thus control gene expression.

C H A P T E R   F O U R

GENE EXPRESSION

Cellular Use of Genetic Information

Overview

A gene is a specific region of DNA, a specific stretch of nucleotides, that contains information for making a particular protein. Cells make thousands of different proteins, and each cell contains thousands of different genes. Gene expression is the process by which the information in DNA is converted to protein.

To make a certain protein, the cell first uses the information in a particular gene to direct the formation of a single-stranded molecule called RNA. The process is called transcription. RNA and DNA are structurally similar, and both use a 4-letter alphabet to store information. The nucleotide sequence of any given RNA molecule is determined by the nucleotide sequence of the gene used to make the RNA. One type of RNA is called messenger RNA, and the information it contains is used to direct the formation of a specific protein by a process called translation. In this process messenger RNA first binds to subcellular workbenches called ribosomes. It then feeds across the ribosomes, and as it does, amino acids, the subunits that make up proteins, align in the order specified by the nucleotides in the messenger RNA. During the alignment process, amino acids are linked together to form a protein chain.

The alignment of amino acids involves a second type of RNA molecule called transfer RNA. It, too, is encoded by genes and is made in the same way as messenger RNA. Transfer RNA molecules serve as adapters, converting the 4-letter alphabet of DNA and RNA into the 20-letter alphabet of proteins. Cells contain at least one transfer RNA for each of the 20 amino acids (letters) that make up proteins. One end of a transfer RNA attaches to a specific type of amino acid, while another part attaches to a specific section of the messenger RNA. Thus the transfer RNAs line up along the messenger RNA in the precise order dictated by the information from the gene. Since each transfer RNA is also attached to one of the amino acids, ordering the transfer RNAs also orders a series of amino acids.

Mechanisms have evolved to control the timing of gene expression, that is, to dictate when a given protein is made from the information stored in its gene. In some cases specific proteins bind to a gene and block its expression. In other cases proteins enhance expression. Since each regulatory protein has its own gene, one gene can influence the expression of another. Sometimes a protein product of a gene influences expression of many other genes, including its own. Elaborate circuits of gene control exist, and they allow a cell to respond quickly to changes in its environment. Successful genetic engineering requires that these regulatory mechanisms be understood.

INTRODUCTION

Gene expression is the process whereby information stored in DNA directs the construction of RNA and proteins. In this process information is first transcribed into RNA, a single-stranded DNA-like molecule. The 4-letter nucleotide code is then translated from RNA to the 20-letter amino acid language of proteins. Gene expression is tightly controlled, and it plays a major role in determining the amount of each type of protein a cell contains. It is the selective control of gene expression that allows multicellular organisms to have different cell and tissue types, since virtually all cells in an organism contain the same genetic information.

The expression of a gene involves many biochemical reactions; consequently, control mechanisms can act during many steps in the process. Specific proteins bind to DNA and specifically raise or lower synthesis of messenger RNA. Other factors, some of them proteins, affect how quickly messenger RNA is broken down into nucleotides. Even binding of ribosomes to the messenger is subject to control. Determining how all of the regulatory factors contribute to the tissue-specific control of human genes is one of the major challenges currently facing molecular biologists.

TRANSCRIPTION

Transcription is the synthesis of an RNA molecule using a DNA molecule as a template; genetic information is converted from a DNA form into an RNA form. RNA is a long, chainlike molecule similar to DNA, but it differs slightly in the sugar part of the backbone: RNA uses the sugar ribose, which is similar to deoxyribose (see Figure 1-4) but has an oxygen attached to the 2′ carbon as well as to the 3′ and 5′ carbons. In addition, RNA is shorter than DNA, and as pointed out earlier, it contains the base uracil (U) instead of thymine (T). In the process of transcription, an enzyme, **RNA polymerase**, recognizes and binds to a DNA nucleotide sequence just before the beginning of the gene. This recognition site, called a **promoter**, positions RNA polymerase properly on the correct DNA strand and points it in the right direction (the strands of DNA are complementary, not identical, and they run in opposite directions). The DNA strands then separate over a short distance, and the polymerase moves into the gene (Figure 4-1). As RNA polymerase moves, it creates a new chain by linking together individual ribonucleotides existing free in the cell. The order of the nucleotides in the new RNA chain is determined by the complementary base pairing rule. If the first letter RNA polymerase encounters in DNA is a T, the enzyme will add an A to the chain it is making. Likewise, if the next DNA letter is G, a C will be added to the new chain. Eventually a stop signal at the end of the gene is reached, RNA polymerase comes off the DNA, and the new RNA chain is released. The new chain is called messenger RNA because it carries information from DNA to sites where the information is used to specify the order of amino acids in proteins. Other RNA molecules, such as transfer RNA and ribosomal RNA, are also encoded by genes, and they are made by the same process as messenger RNA.

TRANSLATION

Messenger RNA transports the information from DNA to subcellular structures called **ribosomes**. Ribosomes are large (in molecular terms) ball-like structures composed of special RNA molecules (ribosomal RNA or rRNA) and ribosomal proteins. It is on them that the informa-

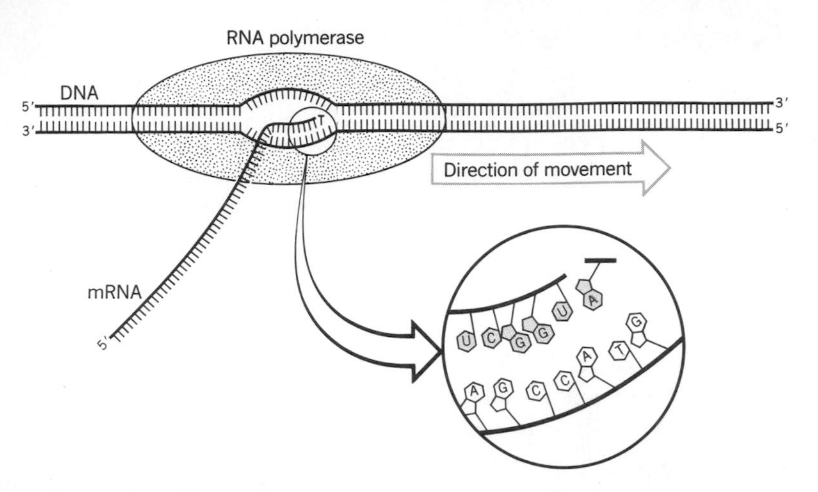

Figure 4-1 Transcription. The enzyme complex called RNA polymerase causes the DNA strands to separate over a short region (10–20 base pairs). The polymerase moves along the DNA, and as it does, it forms an RNA chain using free nucleotides. The order of the nucleotides in RNA is determined by the order of nucleotides in one of the DNA strands by the complementary base-pairing rule. In this example, the nucleotide sequence of the RNA is complementary to that of the lower DNA strand. For simplicity, the DNA strands are not drawn as an interwound helix (see Figure 3-1), nor is complementary base pairing included.

tion in the messenger RNA is **translated** from the nucleotide language into the amino acid language. As a chain of amino acids is made, it spontaneously folds to form the protein specified by the gene in the DNA.

The translation machinery works in the following way. A ribosome attaches to the messenger RNA near a site on the messenger called the start codon, a three-base triplet that indicates where to start reading the message (Figure 4-2). In bacteria, translation begins before messenger RNA synthesis has been completed; thus, both messenger RNA and ribosomes can be attached to DNA simultaneously (Figure 4-2). In a human cell, DNA is located in the **nucleus** while ribosomes are in the **cytoplasm**. Thus the messenger must leave the DNA before attaching to ribosomes. The amino acids which eventually will be linked to form a protein, are normally free in the cell. They are brought to the ribosomes joined to **transfer RNA** (tRNA). To accomplish the

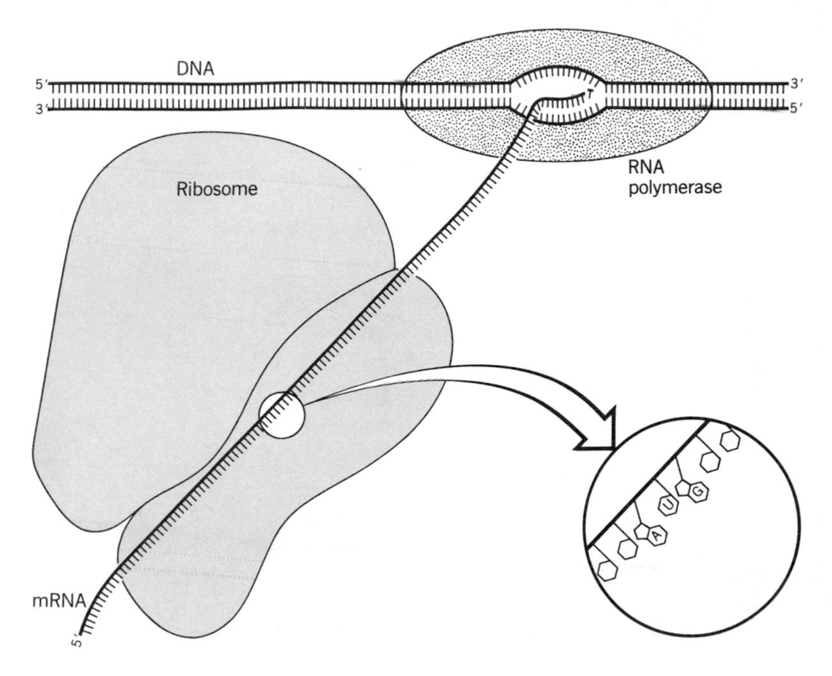

Figure 4-2 Schematic Representation of Messenger RNA Being Formed by RNA Polymerase and Attaching to a Ribosome. The ribosome, composed of two large RNA–protein subunits, binds to messenger RNA. Ribosome–messenger binding requires that a particular transfer RNA also bind to the AUG (or in some instances GUG) codon on the mRNA. This transfer RNA (not shown) is attached to the amino acid destined to become the first in the new protein chain (see Figure 4-4). In bacteria, the messenger RNA is still attached to DNA when it binds to a ribosome. In more advanced organisms, such as humans, the messenger RNA is released from the DNA before attaching to a ribosome.

joining, each type of amino acid is recognized by a special type of enzyme called an **aminoacyl–tRNA synthetase** (Figure 4-3*a*). There are more than 20 different aminoacyl–tRNA synthetases, at least one for each type of amino acid; each of these enzymes recognizes and attaches to only one of the amino acids. Each enzyme is also able to recognize and attach to a specific type of transfer RNA. There is a different type of transfer RNA for each type of amino acid. Once a particular amino acid and a particular transfer RNA have attached to a particular

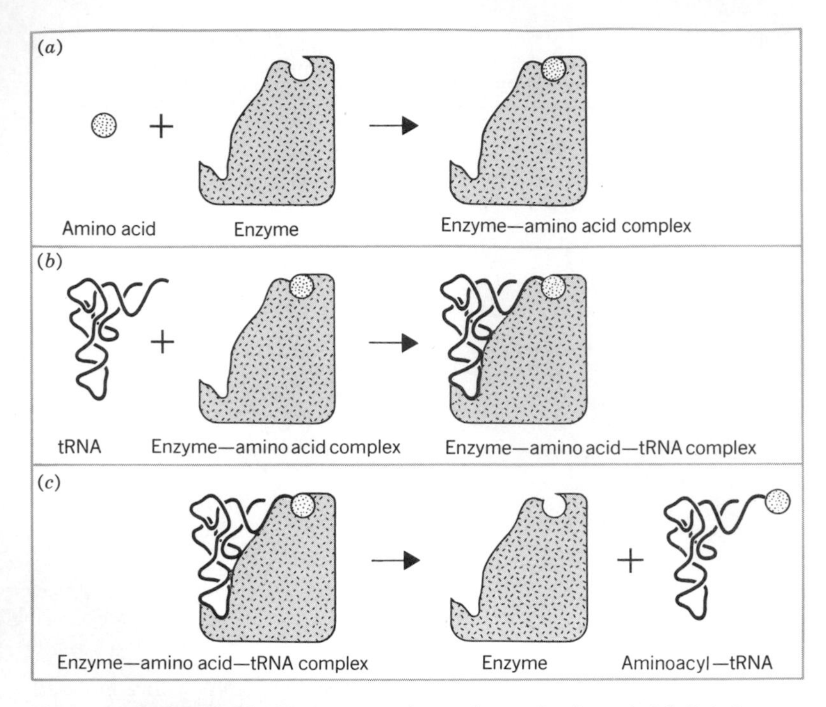

Figure 4-3 Symbolic Representation of an Amino Acid Joining to a Transfer RNA. (**a**) An aminoacyl-tRNA synthetase (labeled "Enzyme") recognizes and attaches to an amino acid. Each of the 20 different amino acids is recognized by a different aminoacyl-tRNA synthetase. (**b**) The enzyme then recognizes and binds to a specific type of transfer RNA (there are more than 20 different types of transfer RNA molecule, at least one for each type of amino acid). In the process, the transfer RNA and the amino acid are joined to form an aminoacyl-tRNA. (**c**) The aminoacyl-tRNA is released from the enzyme, which is then free to repeat the process.

aminoacyl-tRNA synthetase, the synthetase links the amino acid to the transfer RNA (Figure 4-3*b*). The amino acid-transfer RNA pair is then released from the enzyme (Figure 4-3*c*). The net effect is to create an amino acid-RNA pair for each amino acid type.

Each of the transfer RNAs has a sequence of three nucleotides called an **anticodon** (CAU in Figure 4-4). Each type of transfer RNA has a different anticodon. Thus, the particular amino acid at one end of the transfer RNA always corresponds to a specific set of three nucleotides

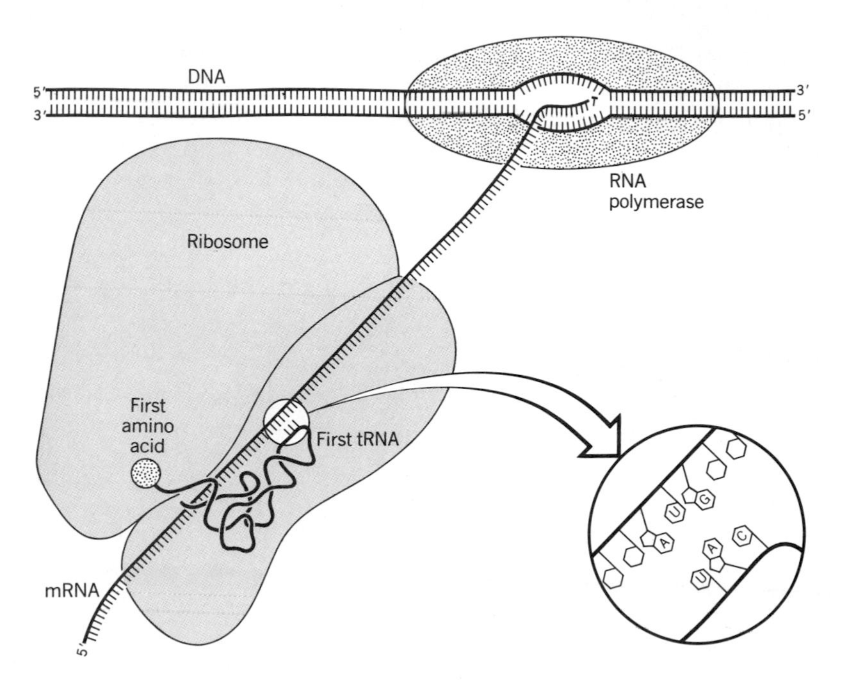

Figure 4-4 Recognition of Codons by Anticodons. The start codon (AUG) on the messenger RNA and the anticodon (CAU) of the first transfer RNA bind on the ribosome. The amino acid destined to be first in the new protein chain is already attached to the first transfer RNA.

in the anticodon region of the transfer RNA. The specificity of the aminoacyl–tRNA synthetase ensures that this is the case. The anticodon on the transfer RNA can be exposed to form base pairs with the messenger RNA (Figure 4-4). One particular transfer RNA has an anticodon triplet complementary to the start codon, or triplet, on the messenger RNA. That transfer RNA and the messenger RNA lock together on the ribosome so that the two triplets, the codon on the messenger and the anticodon on the transfer RNA, form base pairs. This joining is governed by the complementary base pairing rule: if the start codon on the messenger RNA is 5′ AUG 3′, the only transfer RNA that will fit has an anticodon that reads 5′ CAU 3′ (see Figure 4-4; the 5′ and 3′ designations indicate directionality in the two strands, as shown in Figures 1-4 and 3-4). The particular amino acid attached to this transfer RNA is destined to become the first link in the new protein chain.

Figure 4-5 illustrates how the amino acids are ordered in the new protein. The second triplet codon on the messenger is also locked into place on the ribosome next to the first codon. It, too, is recognized by

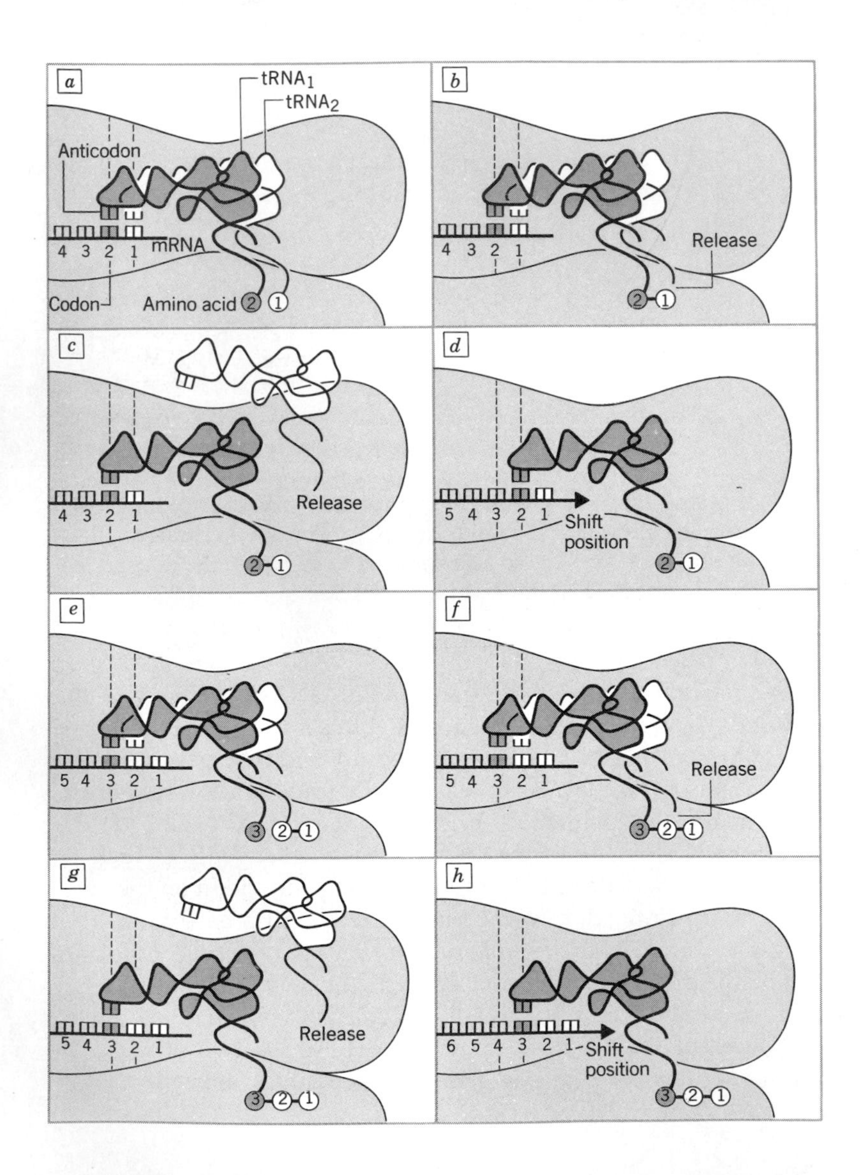

the anticodon of a transfer RNA molecule carrying an amino acid, the amino acid destined to become the second link in the new protein. Other proteins attached to the ribosome then help join the two amino acids together. The first amino acid separates from its transfer RNA, and that transfer RNA separates from the messenger, completing one cycle of the translation process. The messenger now feeds across the ribosome much as a magnetic tape runs over the player head of a tape recorder. One after another the triplet codons are locked into place on the ribosome. The appropriate transfer RNA binds to each triplet, placing the correct amino acid in position to be joined to the growing protein chain. When the stop signal comes along, the messenger falls off the ribosome. The new protein is released into the cell, and it begins to control the specific chemical reaction for which it was designed.

In summary, information from the DNA, encoded by four different letters, is first transcribed into a message (messenger RNA), using a similar 4-letter code. The message then binds to a ribosome. Small RNA molecules, called transfer RNAs, serve as adapters to convert the 4-letter alphabet of DNA and RNA into the 20-letter alphabet of proteins. During protein synthesis, the transfer RNA molecules move amino acids into position along messenger RNA, where it is bound to ribosomes. There the amino acids are linked together to form a protein chain. All organisms on our planet use the same process for making proteins, leading biologists to conclude that life is a continuum.

Figure 4-5 Ordering Amino Acids During Protein Synthesis. **(a)** After the messenger RNA (mRNA), the first aminoacyl–tRNA, and ribosome have formed a complex (Figure 4-4), a second tRNA, with its attached amino acid, is ordered on the ribosome when its anticodon region base-pairs with the second codon of the mRNA. **(b)** Amino acids 1 and 2 are joined; amino acid 1 is released from tRNA 1 (note break). **(c)** tRNA 1 is released from the ribosome. **(d)** mRNA and tRNA 2, now attached to two amino acids, are translocated (shifted over one position on the ribosome). This brings codon 3 into position on the ribosome. **(e)** Aminoacyl–tRNA 3 attaches to the ribosome and forms base pairs with codon 3. **(f)** Amino acid 3 is joined to amino acid 2, repeating step b. **(g)** tRNA 2 is released from the ribosome, repeating step c. **(h)** tRNA 3 and the growing protein chain are translocated, repeating step d. This process continues until a stop codon is reached. At this point the protein chain is released from the last tRNA.

CONTROL OF GENE EXPRESSION: REPRESSION

Organisms have intricate mechanisms for controlling when genes are turned on, that is, when the information in a specific gene will be used to synthesize mRNA and ultimately protein. These mechanisms allow microorganisms to adapt very quickly to changes in the environment. They also allow the cells in higher organisms to develop into a variety of complex structures even though almost all **somatic** (body) cells have exactly the same genetic information. Genetic engineers are trying to determine how cells manage to regulate gene expression. This understanding will then allow the engineers to place genes in bacteria to produce large quantities of a specific protein, into animals or humans to correct defective genes, and into plants to improve food sources.

Biologists have found that nature has a number of ways to regulate gene expression. In **repression**, for example, RNA synthesis is blocked by a particular protein called a **repressor**. The repressor binds specifically to DNA just in front of the gene it controls (Figure 4-6), in a spot called the **operator**. As long as the repressor sits on the DNA, RNA polymerase is unable to start producing a message from that gene. Gene control by repressors has been most thoroughly studied with

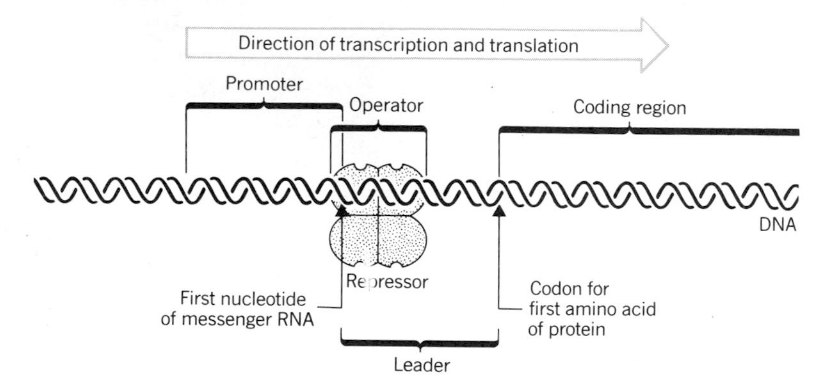

Figure 4-6 Control of Gene Expression by a Repressor. RNA polymerase normally binds to a region of DNA called a promoter. The polymerase then makes a short leader RNA followed by the coding region of the gene. The stretch of DNA that has the information for the RNA leader also serves as a binding site for the repressor. This stretch of DNA is called the operator. When the repressor binds to the operator, it blocks RNA polymerase from binding to this region of DNA and thus prevents synthesis of the messenger RNA.

genes that code for enzymes involved in the breakdown of sugars entering bacterial cells. Unless the particular sugar is present, there is no point in producing the enzyme that breaks it down. If the sugar suddenly becomes available and enters the cell, the repressor binds to the sugar; the repressor–sugar complex is unable to bind to DNA, so repression is lost. As soon as that happens, RNA polymerase binds to the promoter, which in most cases is located near the beginning of the gene. The polymerase then promptly makes the messenger for the enzyme required to break down the sugar. The newly made messenger attaches to ribosomes and is translated, as described earlier in this chapter, producing the degradative enzyme. And as soon as the enzyme has been made, it begins to break down the sugar until none is left for the repressor to bind to. The repressor then attaches to its spot on the DNA, halting production of the messenger for the degradative enzyme. Thus, the cell produces the specialized degradative enzyme only when that particular enzyme is necessary. An important concept is that different repressors regulate different genes by binding to a specific region in front of a gene.

Sometimes several genes in a row are transcribed as a part of a single, long RNA molecule and are controlled by the same repressor; such a unit of genes is called an **operon**. One of the best studied operons is called *lac*, whose three genes are involved in the transport and breakdown of lactose (milk sugar). Figure 4-7 shows how these genes, *lacZ*, *lacY*, and *lacA*, are arranged on the bacterial chromosome. RNA polymerase begins transcribing RNA from the site labeled P in Figure 4-7 and stops after passing through the *lacA* gene. Thus, the information from three genes is transcribed into a single RNA molecule. When the *lac* repressor (encoded by the nearby *lacI* gene, Figure 4-7) binds to the *lac* operator (region O in Figure 4-7), transcription is simultaneously blocked for all three genes. The nucleotide sequence is known for much of this region of the chromosome (part of the operator sequence is shown in Figure 4-7), and molecular biologists are now elucidating the detailed chemistry of repressor–operator interactions.

CONTROL OF GENE EXPRESSION: ATTENUATION

Attenuation is a mechanism for controlling gene expression in which the synthesis of messenger RNA is halted after only a short portion of it has been made. This process has been most extensively studied with

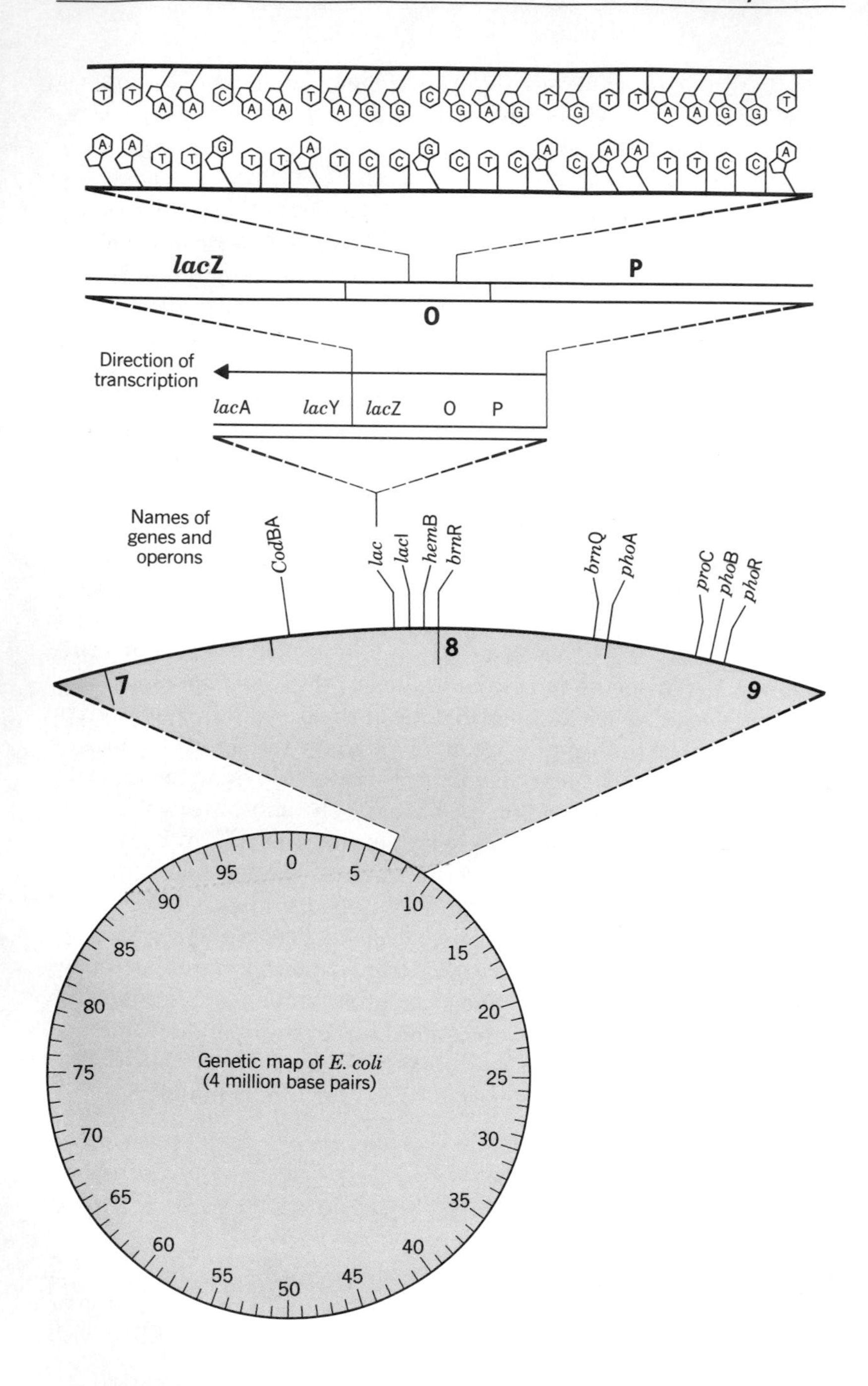
lacZ
P
O
Direction of transcription
lacA
lacY
lacZ
O
P
Names of genes and operons
CodBA
lac
lacI
hemB
brnR
brnQ
phoA
proC
phoB
phoR
7
8
9
Genetic map of E. coli
(4 million base pairs)

genes involved in making amino acids. Amino acids are essential for the health of the cell as building blocks for proteins, so they need to be kept in constant supply. However, their production costs the cell a considerable amount of energy. As a result, mechanisms have evolved that carefully control amino acid production and maintain the correct balance of the 20 different amino acids. Attenuation is one of these mechanisms.

In the attenuation of the **tryptophan** genes, RNA polymerase begins making RNA some distance from the beginning of the first gene in the message, thus creating a **leader RNA** (Figure 4-8). The leader RNA contains a coding region for a short leader protein, and some of the codons in this region specify that tryptophan is to be inserted into the leader protein. Within this leader region is also a variably active stop signal called an attenuator. When tryptophan is abundant, ribosomes are able to translate the leader protein, and as a consequence, the attenuator region of the RNA stops the RNA polymerase by folding into a structure in which two regions of the RNA are held together by complementary base pairing. Consequently, only the short leader message is made. When tryptophan is scarce, the ribosomes stall when they come to the tryptophan codons in the message for the leader protein. The stalled ribosomes sit on the RNA and cause the RNA to fold in a different way, which allows RNA polymerase to continue down the DNA. Thus the entire message is made for the proteins involved in tryptophan production.

Figure 4-7 Organization of an Operon in *E. coli*. The *lacZ*, *lacY*, and *lacA* genes are involved in the metabolism of lactose (milk sugar). All three genes make up the lac operon. They are transcribed into a single messenger RNA molecule. The promoter (**P**) is the region where RNA polymerase binds, the operator (**O**) is the region where the repressor binds, and *lacZ* is the gene for the enzyme that degrades lactose. The exact sequence of DNA nucleotides is known for the lactose gene region; the nucleotide sequence for the binding of the lactose repressor protein to the DNA and the relationship of this stretch of DNA to the chromosome are shown. An *E. coli* DNA molecule is a large circle that, if stretched out, would be 1000 times longer than an *E. coli* cell. The genes in this DNA molecule are arranged in a circular map divided into 100 units. An enlargement of the region of the map near position 8 shows the location of the lac operon. For clarity, the DNA strands are not shown as a double helix, and hydrogen bonds between bases are omitted.

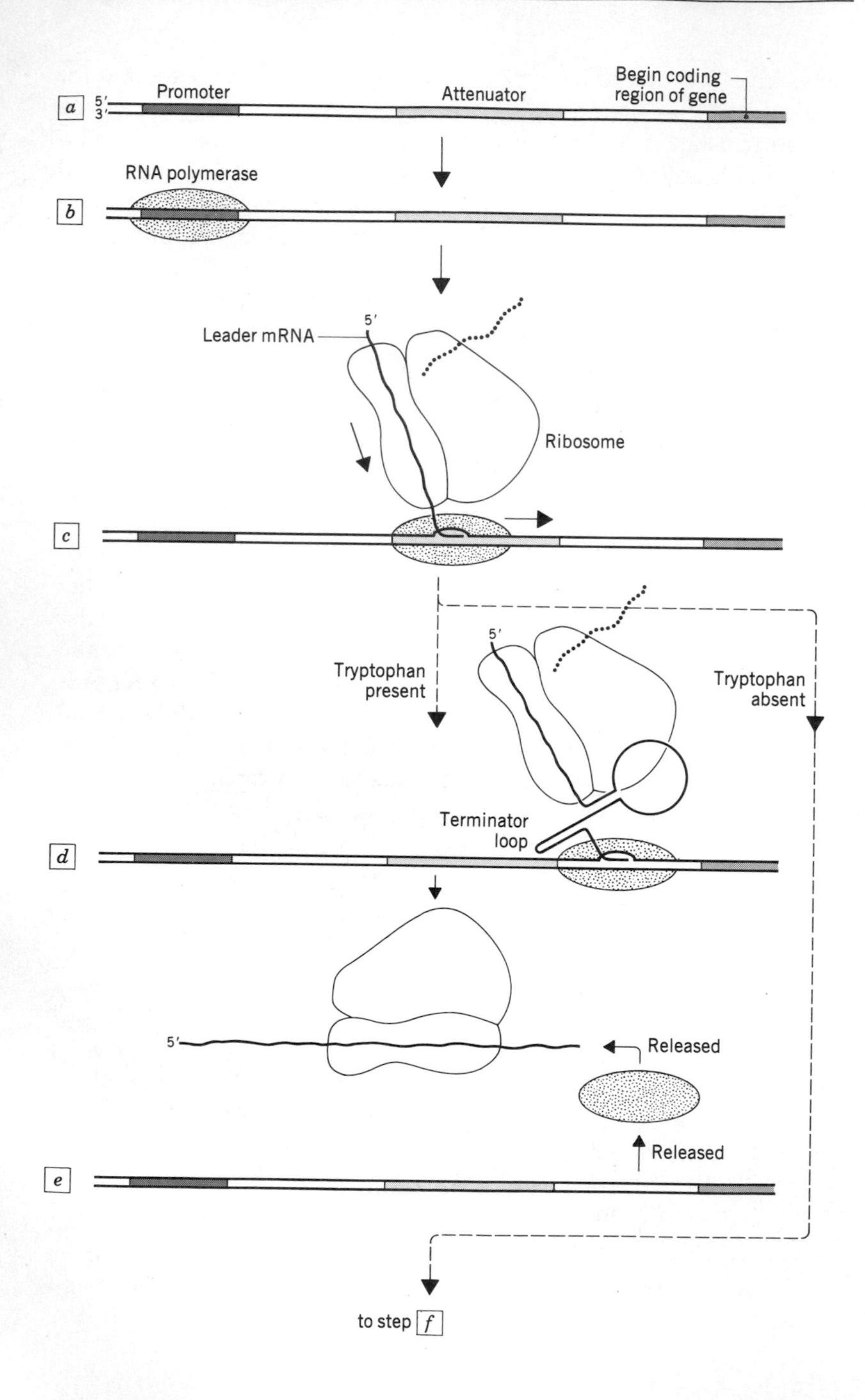
a
5′
3′
Promoter
Attenuator
Begin coding region of gene
b
RNA polymerase
Leader mRNA
5′
Ribosome
c
5′
Tryptophan present
Tryptophan absent
Terminator loop
d
5′
Released
Released
e
to step f

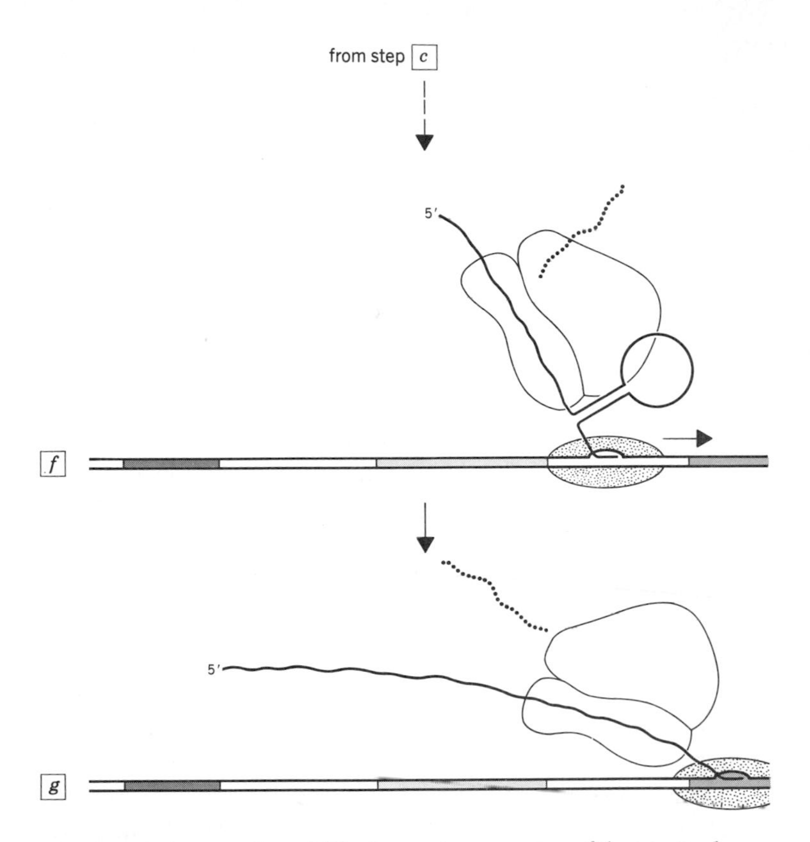

Figure 4-8 Attenuation. (**a**) In the upstream region of the tryptophan operon, the promoter is about 200 nucleotides from the beginning of the coding region of the first gene in the operon. (**b**) RNA polymerase binds to the promoter. (**c**) Transcription of the leader mRNA begins, ribosomes bind to the mRNA, and synthesis of the leader peptide begins. (**d**) In the presence of tryptophan, the ribosome moves along the leader RNA and a loop in RNA forms that acts as a transcription terminator. (**e**) Transcription stops, RNA polymerase is released from the DNA, and the mRNA is released from the polymerase–DNA complex. (**f**) In the absence of tryptophan, the ribosome stalls at a pair of tryptophan codons because there is no tryptophan to incorporate into the leader peptide. This allows a specific loop to form in the RNA that prevents the formation of the terminator loop. (**g**) RNA polymerase continues to travel down the DNA, synthesizing messenger. The first gene of the operon has its own ribosome binding site, so new ribosomes can bind to translate the message from the gene.

CONTROL OF GENE EXPRESSION: ACTIVATION

A general strategy for turning on genes involves the binding of special proteins near promoters to facilitate the proper binding of RNA polymerase. Gene activation appears to be particularly widespread among genes in complex organisms where large numbers of related genes must be regulated together to ensure that specific cell types will develop at the correct time.

In higher cells, cases involving several different proteins and several regions of DNA in gene activation have been found. Usually these control regions lie upstream from the start of the coding region of the gene, and often multiple regions are sites for the binding of specific proteins. Some of these proteins are general activators; that is, they are found in many cell types. Others are found only in certain cell types, such as liver or kidney. When the correct constellation of control proteins is present, the gene becomes activated and its protein product is made. Of course the control proteins are also produced by genes, and the regulatory networks can be quite complex. Cases have been found in which a regulatory protein activates its own gene as well as other genes. Thus once the regulatory protein has been made in a particular type of tissue, its self-activation would assure permanent activation for the life of the organism. Such a phenomenon could contribute to the differentiation of our cells into specific types.

DNA regions important for the control of a gene can be located far from the gene, and small segments called **enhancers** can activate genes even when the enhancer is a thousand nucleotides from the gene. Some enhancers are even located downstream from a gene or in the middle of it. Enhancers appear to be binding sites for specific proteins that in turn bind to other proteins attached to regions of DNA near the gene. In these cases the DNA must loop to allow the various proteins to bind to each other and stimulate RNA polymerase action. A scheme for enhancer action is shown in Figure 4-9.

PERSPECTIVE

During the past four decades a great deal has been learned about DNA structure and gene expression. It is probably useful to place some of these concepts into perspective and to point out their relevance to genetic engineering.

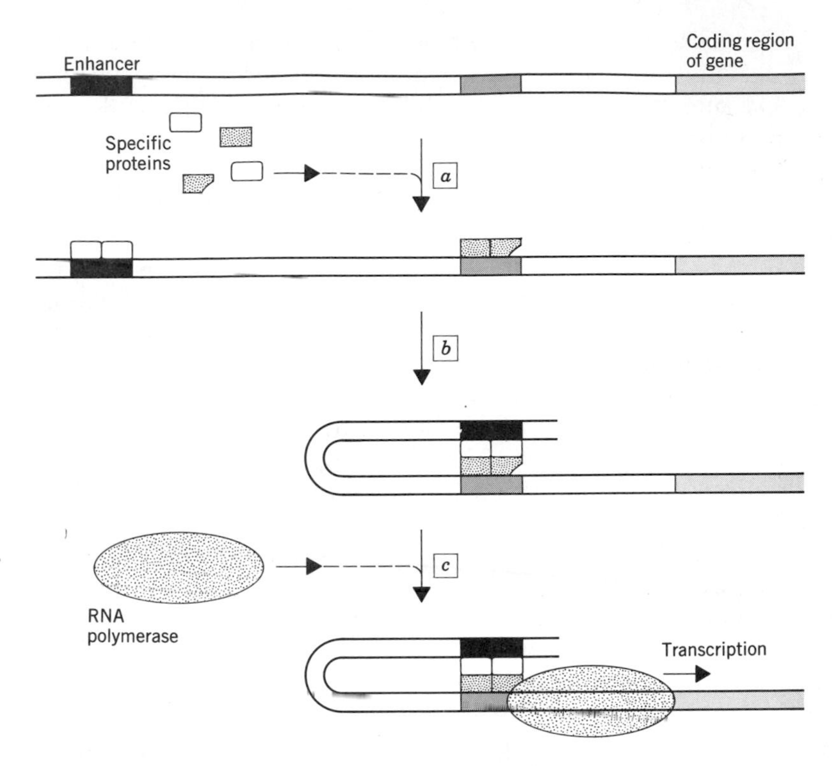

Figure 4-9 Gene Activation Involving an Enhancer. (**a**) Specific proteins bind to regions upstream from the gene. (**b**) The two sets of proteins bind to form a complex. (**c**) RNA polymerase binds to the complex.

1. Living cells are composed of molecules, many of which are constantly being converted into other molecules through a series of steps called metabolism.
2. These chemical conversions are carried out and controlled by protein molecules called enzymes. Usually a different enzyme is responsible for each step in a conversion pathway. Thus, to alter the characteristics of an organism (i.e., to change the chemical reactions), one must change the proteins controlling these reactions. The protein content of an organism can be changed temporarily by injections (as in the case of insulin) or permanently by genetic engineering.

3. Genetic engineering is based on the observation that the information required to make each protein is stored in a long, chainlike molecule called DNA. This information is arranged into short regions called genes, such that each gene contains the information for the production of one protein. Genetic engineers change a particular protein by changing the information in DNA. Once the change in DNA has occurred, every protein made from the new information is of that new variety. The change is permanent because DNA reproduces itself every time a cell divides.

Gene cloning is a form of genetic engineering. We can review how genes are cloned by adding details to the scheme sketched earlier. First, DNA is removed from a donor organism and cut with enzymes to make single genes accessible. The DNA fragments are inserted into small, infectious DNA molecules (cloning vehicles), which are used to carry the DNA into microorganisms. The details of cutting and inserting as well as the biology of cloning vehicles are described in later chapters. Unfortunately, the cutting enzymes produce thousands of DNA fragments. One of the major tasks of genetic engineers is to find the particular gene for the protein they wish to change. By putting the DNA fragments into bacteria, it is possible to locate a specific piece of DNA. First, the fragments are transferred into bacterial cells so that each cell receives only one fragment. Each cell multiplies to form a colony; at the same time many copies of the DNA fragments are made. Every colony is then tested for the presence of the DNA fragment being sought. One testing procedure is based on the chemical structures of DNA and RNA and the principle that complementary base pairs tend to stick together. Since messenger RNA from a specific gene is complementary to one of the DNA strands of that gene, it can form a double-stranded structure with DNA. Consequently, messenger RNA can in principle be used to test bacterial colonies for the presence of a specific gene by looking for RNA : DNA hybrids, double-stranded nucleic acid molecules in which one strand is RNA and the other is DNA (hybrid formation is described in the next chapter).

Once a bacterial colony that contains the gene being sought has been identified, the gene cloner must recover the gene from the bacterial colony. This is accomplished by removing the DNA from the bacterial cells and purifying the cloning vehicle that carries the gene. The gene can then be cut out of the cloning vehicle, inserted behind bacterial regulatory regions, reintroduced into bacterial cells, and cultured for large-scale production of the protein specified by the gene. Alternatively, the gene can be inserted into a different cloning vehicle to carry it into cultured animal cells, which in turn may be put into whole animals. Although the latter scenario has been accomplished, much more information about the regulation of genes in animals must be obtained before such techniques can become generally useful.

Questions for Discussion

1. Describe how information flows, in terms of molecules, when gene expression occurs.
2. Although it seems most efficient to control gene expression by regulating when RNA synthesis occurs, sometimes control is exerted by regulating when translation occurs. Frog eggs represent one example. In this case messenger RNA accumulates within the egg cell, and translation is delayed until fertilization by a sperm has occurred. How might this accumulation of RNA benefit the developing frog embryo?
3. Cases have been found in which a single type of repressor binds to operators of many genes that are widely separated on a chromosome. Inactivation of the repressor then leads to transcription from the whole set of genes. If the repressor gene is very active, removal of the inducer will then lead to rapid repression of all the genes in the system. But how might the repressor gene be controlled to prevent the cell from filling up with repressor, hence lessening the sensitivity of the response to inducer?
4. RNA polymerase is composed of several separate proteins called subunits. In bacteria one of these subunits is called **sigma** (σ). Sigma is thought to be involved in the recognition of promoters. The preference of RNA polymerase for promoters changes when the sigma subunit associated with the enzyme changes. By controlling the production of certain sigma subunits, it is possible to control the expression of large sets of genes. Under what conditions would this sigma subunit mode of regulation be more suitable than the repressor system described in question 3? Among the factors to consider are response time, ability to amplify the effect of the stimulus, and reversibility.

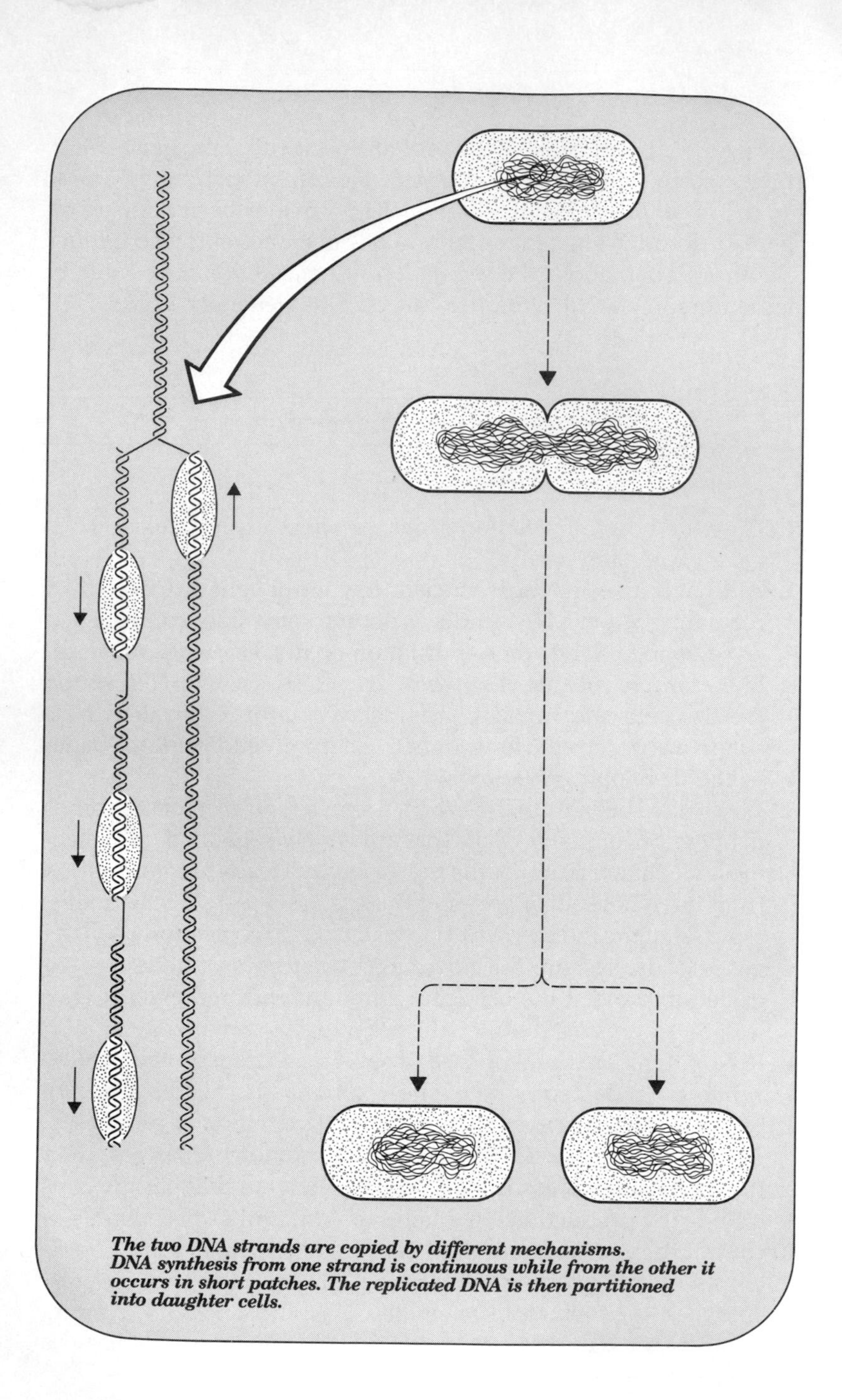

The two DNA strands are copied by different mechanisms. DNA synthesis from one strand is continuous while from the other it occurs in short patches. The replicated DNA is then partitioned into daughter cells.

C H A P T E R   F I V E

REPRODUCING DNA

Information Transfer from One Generation to the Next

Overview

DNA is copied by a group of proteins traveling together along a double-stranded DNA molecule. As the proteins move along the DNA, they separate the two DNA strands and make a new strand adjacent to each old one; one double-stranded molecule becomes two. Each DNA has one old and one new strand, and the nucleotides in these two strands are again complementary. Occasionally errors occur while the new strands are being made, and the nucleotide sequence in the new strand is not perfectly complementary to that in the old strand. If not corrected, these errors, called mutations, are passed faithfully from one generation to the next.

The two strands in a DNA molecule are not replicated in exactly the same way because they run in opposite directions. Replication of one old strand occurs by formation of one long, continuous complementary strand. Replication of the other, however, occurs by formation of short patches, which are later connected by an enzyme called DNA ligase. Enzymes involved in DNA replication have been purified, and some are important tools for gene cloning. For example, DNA ligase is used to join DNA fragments together. One form of DNA polymerase, an enzyme that copies the old strand into a new one, is used to make highly radioactive DNA in test tubes. This radioactive DNA is utilized to find bacterial colonies that contain specific cloned genes.

INTRODUCTION

Earlier we defined life as a set of chemical reactions organized in such a way that this set can reproduce itself. The preceding chapters touched on how DNA and enzymes organize the reactions. Now, to introduce many of the tools used in gene cloning, the aspect of reproduction is examined. For the present discussion, reproduction is defined as the duplication of the information content of the cell, followed by segregation of this information into two newly formed daughter cells. Since information is stored in DNA, knowing how DNA duplicates is crucial to understanding reproduction. The next section discusses the machinery responsible for **DNA replication**, the technical term for reproduction of DNA. Understanding mutations has been important for deciphering how DNA functions, so the chapter continues with a section on DNA structure and mutation. The third section outlines how enzymes are obtained and handled, since enzymes involved in DNA replication are important for manipulating DNA. The following section discusses how radioactive probes are used to locate bacterial colonies containing cloned genes, and the final section outlines a method for amplifying DNA.

DNA REPLICATION

To provide genetic continuity from one generation to the next, DNA not only must be chemically stable, but it also must be copied accurately during replication. If this were not the case, the DNA of an offspring would contain information different from that of its parents. Then the proteins in the offspring would differ from those in the parents, and the characteristics of the two generations would no longer be the same. Accurate copying is accomplished in the following way. The two strands of DNA separate, permitting each to act as a template for formation of a new strand. Nucleotides are aligned along each DNA strand according to the complementary base pairing rule (Figure 3-5), and they are joined to form a new DNA strand. Two DNA molecules arise from one, and they contain identical information (Figure 5-1). The two DNA molecules move to different parts of the cell, cell division occurs between the DNA molecules, and two daughter cells arise having identical DNAs (Figure 5-2).

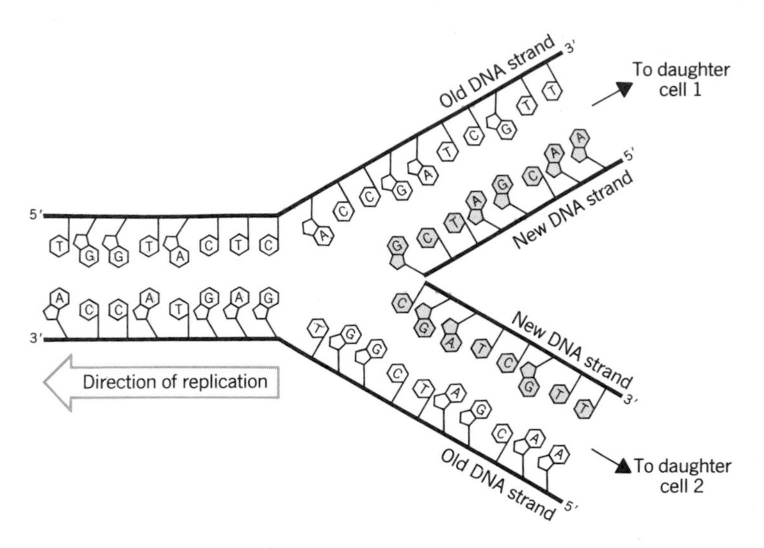

Figure 5-1 Two DNA Molecules Arise from One. Base-pairing complementarity allows information to be copied exactly. Notice that each daughter cell will receive a DNA molecule having exactly the same nucleotide sequence. For clarity, the strands are not drawn as an interwound helix (see Figure 3-1), and hydrogen bonding between complementary base pairs (Figures 3-3 and 3-5) is omitted.

Since biochemists can replicate DNA in test tubes by adding a small number of purified components, much is known about the replication machinery. The process can be divided into a number of steps. First, specific proteins form a complex with DNA at a specific site in the DNA called an origin of replication. Next, the two DNA strands begin to unwind, producing a **replication fork** that moves through the double-stranded DNA molecule as replication occurs. The process is much like unzipping a zipper (Figure 5-1). Thus two single strands are created, exposing the bases. An enzyme called **DNA polymerase** binds to one of the single strands and moves in the direction of fork movement, closely following the zipper (Figure 5-3). As it moves along the single strand, DNA polymerase mediates formation of base pairs between free nucleotides (links not yet in a chain) and the linked DNA nucleotides. The alignment obeys the complementary base pairing rule, so wherever an A occurs in the single DNA strand, a T is aligned opposite it. As soon as a free nucleotide has lined up at the end of the

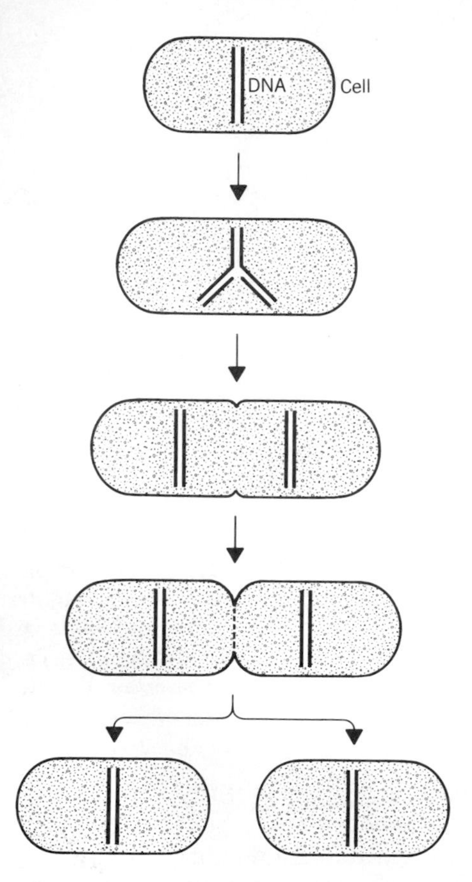

Figure 5-2 Replication and Segregation of Bacterial DNA. While the cell is growing larger, its DNA replicates. The two daughter DNA molecules move to opposite parts of the cell. A new cell wall forms between the DNA molecules, and two daughter cells are produced. A similar process occurs in the cells of higher organisms. However, they often have many DNA molecules, so the sorting system is more complex.

growing chain (Figure 5-3), and opposite to its complement in the template strand, DNA polymerase links it to the new chain. The polymerase moves down the chain one position, aligns the next nucleotide, and links it to the growing chain. Thus a new single-stranded DNA chain is formed using the information contained in the old one. As the new chain forms, it is already base-paired with the old one; thus a dou-

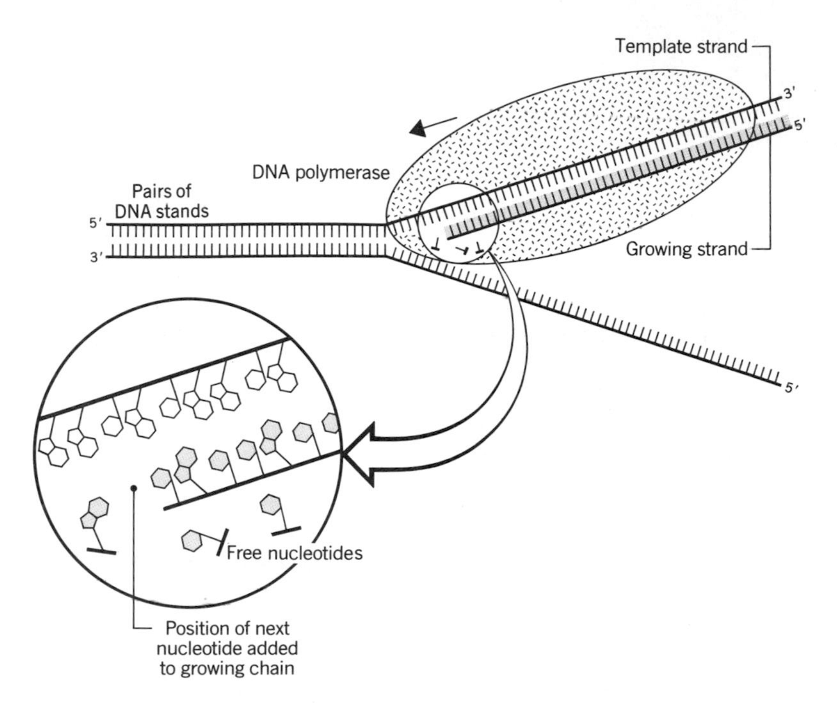

Figure 5-3 Function of DNA Polymerase. The DNA strands separate as DNA polymerase and other proteins bind to the DNA. In the replication of one strand, DNA polymerase follows the replication fork, forming a new strand whose nucleotide sequence is complementary to that of the old, template strand. Nucleotides are added to the end of the growing chain one at a time. The order of the nucleotides in the growing chain is determined by the order in the template strand. For clarity, the DNA is drawn as two parallel strands rather than as a double helix as in Figure 3-1, and hydrogen bonding between base pairs (Figure 3-3) is omitted.

ble-stranded DNA molecule containing one new strand and one old one is produced.

A particularly interesting detail in the scheme just described is that DNA polymerase always travels in the same direction along a DNA chain (5′ to 3′ in the growing strand). Like a motion picture film, DNA has directionality. The nucleotides are bound much like a chain of elephants hooked trunk to tail, and the polymerase recognizes this directionality. This aspect of structure is important because the two strands in a double-stranded DNA molecule run in opposite directions (see Figure 3-4*b*). As the replication fork moves through the DNA molecule, it unzips the DNA and makes single strands available for replication. DNA polymerase on one strand will follow the fork (Figure 5-4), continuing in the same direction as the fork. Polymerase on the other strand, however, must move in the opposite direction, away from the fork. Synthesis opposite to fork movement occurs in short patches, which in bacteria are about 1000 nucleotides long (see frontispiece, Chapter 5). Once the polymerase has made one patch, we could imagine that it leaps back toward the vee in the fork and begins making another piece of DNA until it runs into the patch of new DNA it had just laid down. Then the polymerase must again catch up with the moving fork. Whether the polymerase actually leaps toward the moving fork is not known; molecular biologists are currently testing ideas that don't require the polymerase to leap. An important point to emerge from this type of study is that DNA polymerase is unable to join the two patches of new DNA together. Another enzyme, called **DNA ligase**, is required to connect the patches. Ligase is now an important tool genetic engineers use to join DNA fragments together.

DNA STRUCTURE AND MUTATIONS

Occasionally errors are made during DNA replication, and the errors may be passed on to the next generation of cells. Changes in genetic information can have serious consequences. Examination of the genetic code (Figure 3-6) reveals how specific nucleotide changes in DNA can lead to specific amino acid changes in protein. An example is illustrated in Figure 5-5. In an offspring a single nucleotide pair might be changed in the gene; for example, an A-T pair might be converted to a G-C, C-G, or T-A pair. Consequently, an incorrect amino acid may be inserted into the protein specified by the gene containing the error. An incorrect amino acid will often change the structure of the protein enough to make it inactive. Since inactive proteins frequently alter the chemistry of cells and organisms, it would not be surprising to find that the offspring looked different from the parent or acted in different ways. Such an altered organism is called a **mutant**, and if the mutation (i.e., the nucleotide change in the DNA) has produced an inactive protein in an essential gene, the mutant organism will die. Occasionally mutation produces a better protein in the offspring, making it better able to survive and reproduce in that particular environment. In such

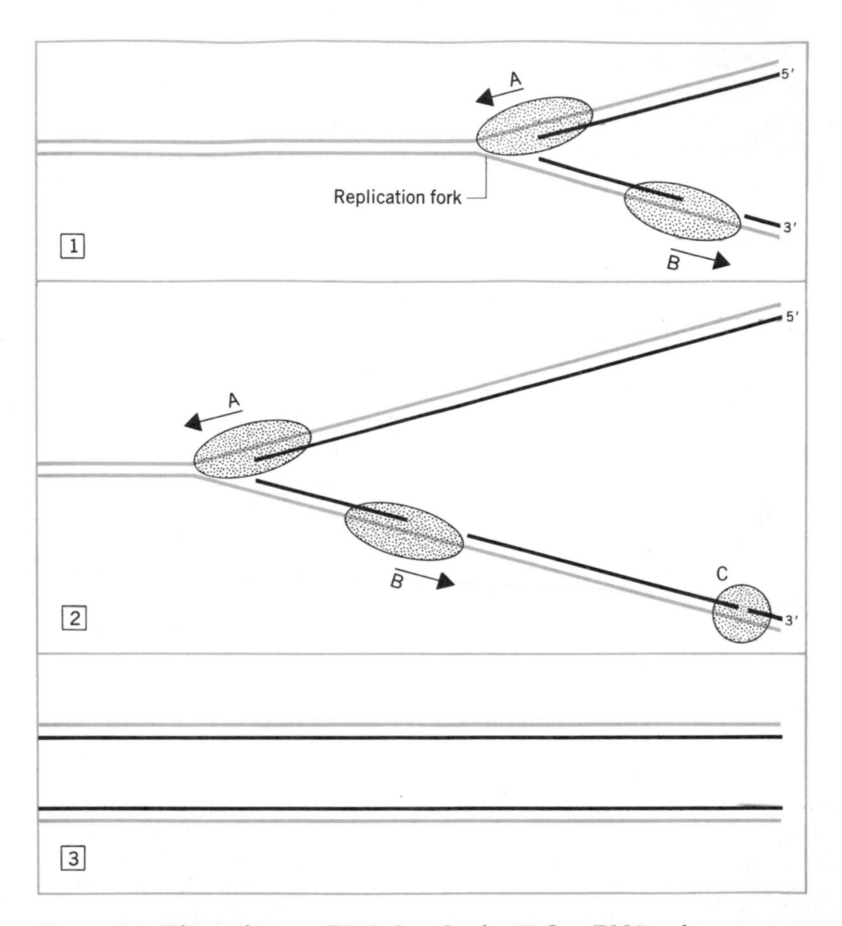

Figure 5-4 Discontinuous DNA Synthesis. **(1)** One DNA polymerase complex (**A**) moves continuously along an old single strand (shaded) synthesizing a new strand (solid) in the direction of replication fork movement while the other (**B**) moves in the opposite direction. **(2)** Polymerase **B** synthesizes short patches of DNA. Small gaps are filled in by another type of DNA polymerase, and the DNA fragments are joined by DNA ligase (**C**). **(3)** The result is two double-stranded DNA molecules.

cases the mutant may eventually become the dominant type of organism in the population, and a small step in evolution will have occurred.

A number of different changes in DNA have been discovered that give rise to mutations (Figure 5-6). In addition to the example given

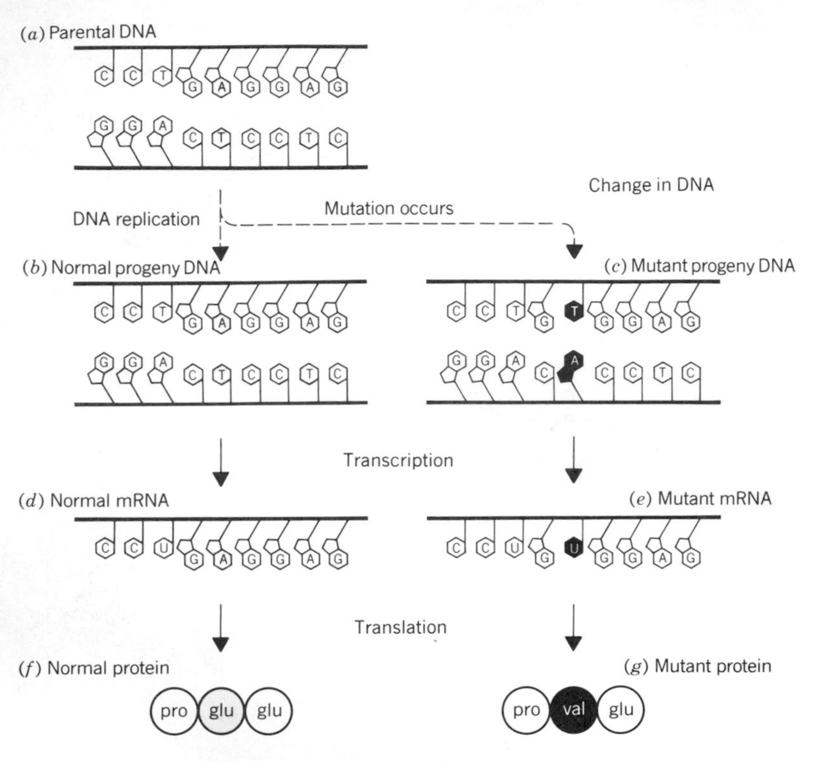

Figure 5-5 A Point Mutation Changes the Sequence of Amino Acids in a Protein. DNA replication is very accurate, so the nucleotide sequence in the progeny DNA (**b**) is identical to that of normal parental DNA (**a**). Occasionally an error is made. In this example, a particular A-T base pair in parental DNA changes to a T-A pair in the mutant progeny DNA (**c**). During transcription the information in DNA is converted into messenger RNA. The mutation in DNA results in a conversion of particular GAG codon in normal messenger RNA (**d**) into a GUG codon in mutant messenger RNA (**e**). During translation of the information into protein, GAG codes for the amino acid glutamic acid (glu) (**f**), while GUG codes for valine (val) (**g**) (see Figure 3-6). The two amino acids have very different chemical properties. Since the structure of the resulting protein is determined by the precise order of the amino acids, the mutant protein will differ significantly from the normal protein. The differences between the normal and mutant molecules shown are identical to those found between healthy people and patients suffering from sickle-cell anemia.

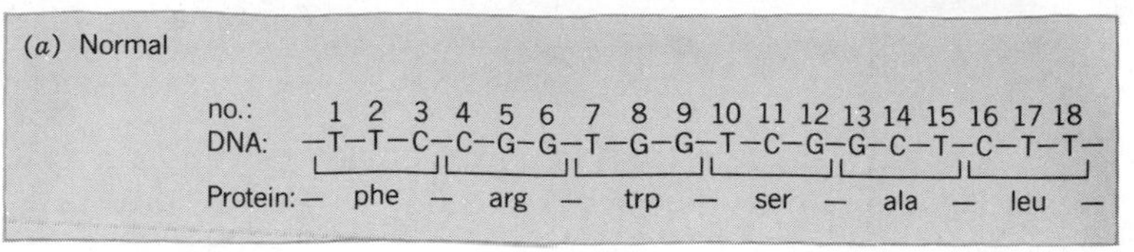

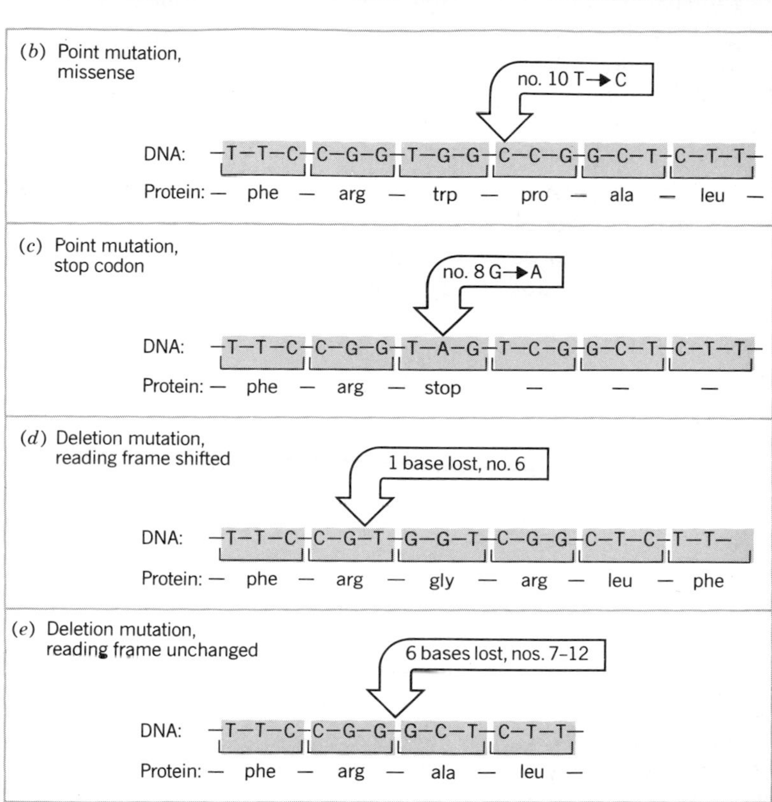

Figure 5-6 Common Types of Mutation. (**a**) A normal nucleotide sequence for one strand of DNA codes for a protein having the six amino acids listed (phe, arg, trp, etc). The codons for the amino acids are bracketed above the respective amino acids. (**b**) If a T is changed to a C (arrow), the resulting mutant protein has a proline where serine is normally located. This type of change is called a ***missense*** mutation. (**c**) If a G is changed to an A (arrow), a stop codon is created and protein synthesis halts. This change is called a nonsense mutation, and the mutant protein is shorter than the normal protein. (**d**) Deletion of one base throws the reading frame out of register (frameshift mutation), and incorrect amino acids (gly, arg, leu) occur in the mutant protein. (**e**) Removal of six bases produces a deletion mutation and a protein missing two internal amino acids (trp, ser).

above, changing a single nucleotide occasionally converts a normal triplet codon into a stop codon (see Figure 3-6) — then the operation of the protein synthesis machinery prematurely terminates (Figure 5-6*c*). If such a mutation occurred near the beginning of the gene, only a small fragment of the protein could be made, with obviously serious effects. In another kind of mutation a nucleotide is lost during replication; that is, the replication machinery skips a letter (Figure 5-6*d*). In this case the information is thrown out of the correct reading frame, and many incorrect amino acids are placed in the protein. This is called a **frameshift** mutation. In still other cases, a large stretch of DNA is lost (Figure 5-6*e*). Fortunately, mutations rarely occur in the absence of chemical agents, and as a result organisms have stable characteristics.

The accuracy of the replication machinery is very impressive. In bacteria, detectable mutations (errors) in a given gene arise at a frequency of only one in a million cells per generation, even though the replication apparatus copies DNA at a rate of 50,000 base pairs per minute. The key to understanding this process lies in the principle of base-pairing complementarity. The important point is that biological molecules recognize each other by fitting together like locks and keys. Nearly perfect fits are required for two molecules to bind together, and when the fit is good, the binding can be strong. Thus, as long as the rule illustrated in Figure 3-5 is obeyed (namely, that A bind only to T and G only to C), there will be no errors and replication will proceed properly.

Within this framework, any chemical that converts one letter to another in the old DNA strand is capable of "tricking" the replication machinery into inserting the wrong nucleotide into the new DNA strand as it is made. Chemicals that alter the information in DNA are called **mutagens**. Chemical mutagens, which are everywhere in our environment, and physical factors such as **ultraviolet light**, have become major threats to human health. There is little doubt that both kinds of mutagen can lead to certain types of cancer. Bacteria are now being used to test household items, food additives, and pesticides for their ability to cause mutations and presumably cancer. The test is based on our ability to easily detect the creation of bacterial mutants by observing whether they grow on agar plates containing particular nutrients.

HANDLING ENZYMES

Thus far enzymes have been considered to be tiny agents that somehow join nucleotides to form DNA or RNA chains, join amino acids to form protein chains, and in general direct the chemical reactions of the cell. To provide a better understanding of enzymes, the processes they control, and how they are used in genetic engineering, it is necessary to digress from describing biological principles into a discussion of some of the technical aspects of handling enzymes. Methods for obtaining DNA polymerase are outlined below to illustrate how enzymes are detected and purified.

To be useful for gene cloning, enzymes must be removed from living cells. The first step of purification is to obtain a large batch of cells, about a pound of cells in the case of bacteria. Then the cell walls must be broken to liberate the enzymes. This can be done in a number of ways; one method is to combine the cells with tiny glass beads and grind the mixture in a blender. The mixture of cellular components is called a **cell extract**. Then the enzyme of interest is separated from other cell components by a series of physical manipulations. Separation is possible because every type of enzyme is physically and chemically different from every other type of molecule in the cell. For example, DNA polymerase is relatively small compared to many cell components. Thus one purification step might be to allow the large cellular debris to settle to the bottom of a test tube. The settling process is greatly accelerated by centrifugation (Figure 5-7). DNA polymerase is expected to remain in the fluid at the top of the tube. A crucial part of this purification step is determining whether DNA polymerase settled to the bottom of the tube or stayed in the fluid. Thus an **assay** for the enzyme—that is, a way to detect and follow the enzyme through a series of purification steps—must be developed.

To develop an assay for a particular enzyme, a biochemist must understand the nature of the chemical reaction stimulated by the enzyme. From the discussion in the first part of this chapter, one could imagine that DNA polymerase, using a single-stranded DNA template, links free nucleotides together to form a DNA chain. Figure 5-8 shows how this concept can be used to determine whether an extract of broken cells contains DNA polymerase. Generally the number of free nucleotides that are linked together in a test tube reaction is very

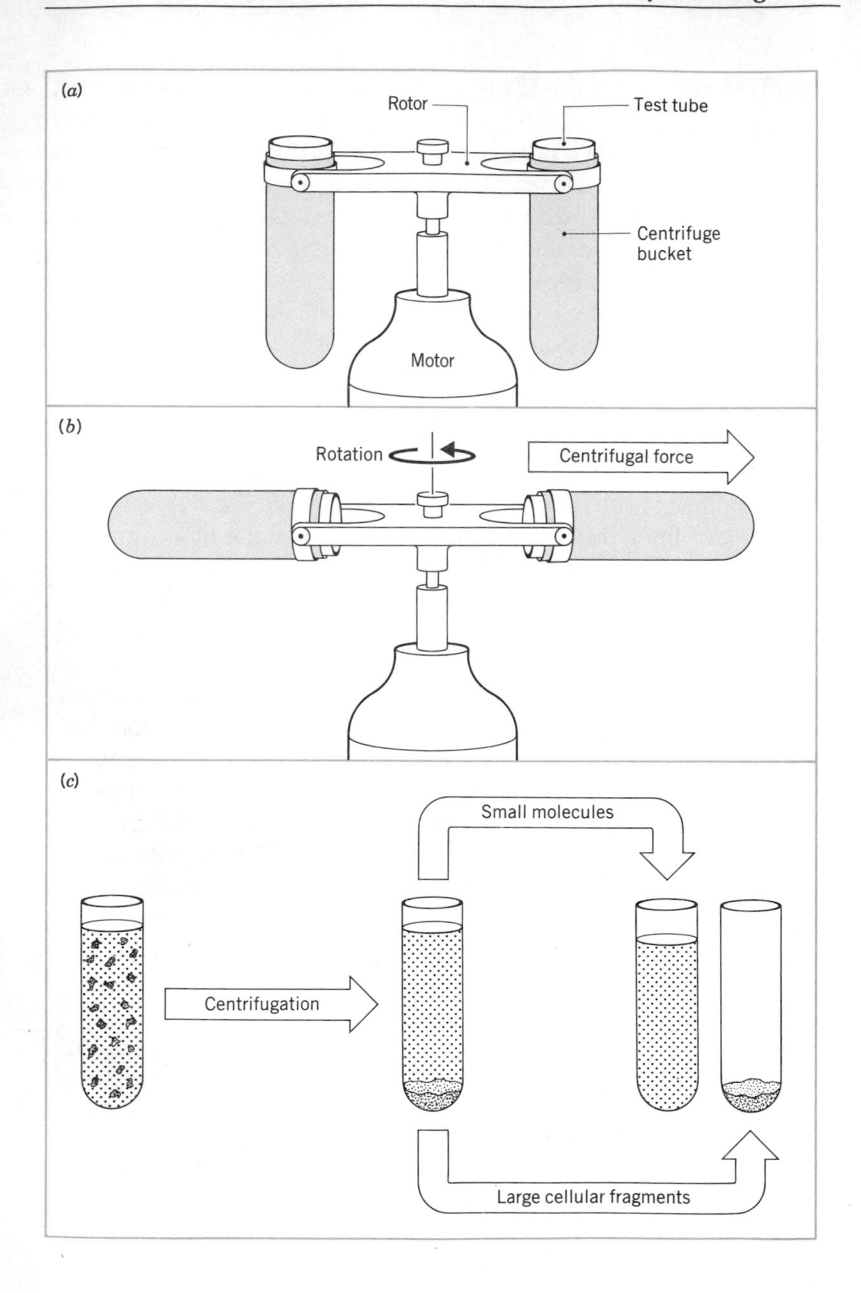
(a)
Rotor
Test tube
Centrifuge bucket
Motor
(b)
Rotation
Centrifugal force
(c)
Small molecules
Centrifugation
Large cellular fragments

small; therefore, it is necessary to have a very sensitive measure for this conversion. Radioactive isotopes are used for making such measurements because very small amounts can be detected. As mentioned earlier, radioactive isotopes are unstable forms of atoms that spontaneously disintegrate. Those used in biochemical studies emit high energy particles that can be detected on film or with instruments such as Geiger counters. A wide variety of radioactive molecules are commercially available.

Operationally, a mixture is prepared consisting of a cell extract that contains the DNA polymerase, the four nucleotide triphosphates, one of which is radioactive, and a single-stranded DNA template to which is attached a short complementary **oligonucleotide** called a **primer** (see Figure 5-8*a*). DNA polymerase cannot begin DNA synthesis with only a template strand; it must always add nucleotides to the end of a preexisting strand. After the mixture has been incubated for an hour or so at 37°C (body temperature), cold acid is added to the mixture. DNA molecules clump together to form a white, stringy **precipitate**, a solid that settles to the bottom of the test tube. If the acid-containing mixture is then poured into a funnel lined with filter paper, the solution will pass through the filter paper, but precipitated DNA will stick to it. Free nucleotides are relatively small, and they do not precipitate when acid is added to the mixture; consequently, they are not trapped by the filter paper. The only radioactive molecules that can stick to the filter paper are the nucleotides that have become a part of the DNA. Thus, the amount of radioactivity on the filter paper is a measure of the number of the nucleotides that have been linked to form DNA, which in turn is a measure of the DNA polymerase activity present.

Figure 5-7 Fractionation of a Cell Extract by Centrifugation. (**a**) Schematic diagram of a swinging bucket rotor. Test tubes filled with the cell extract are placed in the centrifuge buckets. (**b**) The motor causes the rotor to spin, the buckets swing into a horizontal position, and the molecules in the cell extract sediment (migrate) toward the bottom of the test tube. Ultracentrifuges can generate forces in excess of 500,000 times gravity. (**c**) The force generated by the centrifuge causes molecules to separate on the basis of size and shape. After the large molecules have sedimented to the bottom of the tube, the upper solution can be carefully removed with a **pipette** and placed in a second tube.

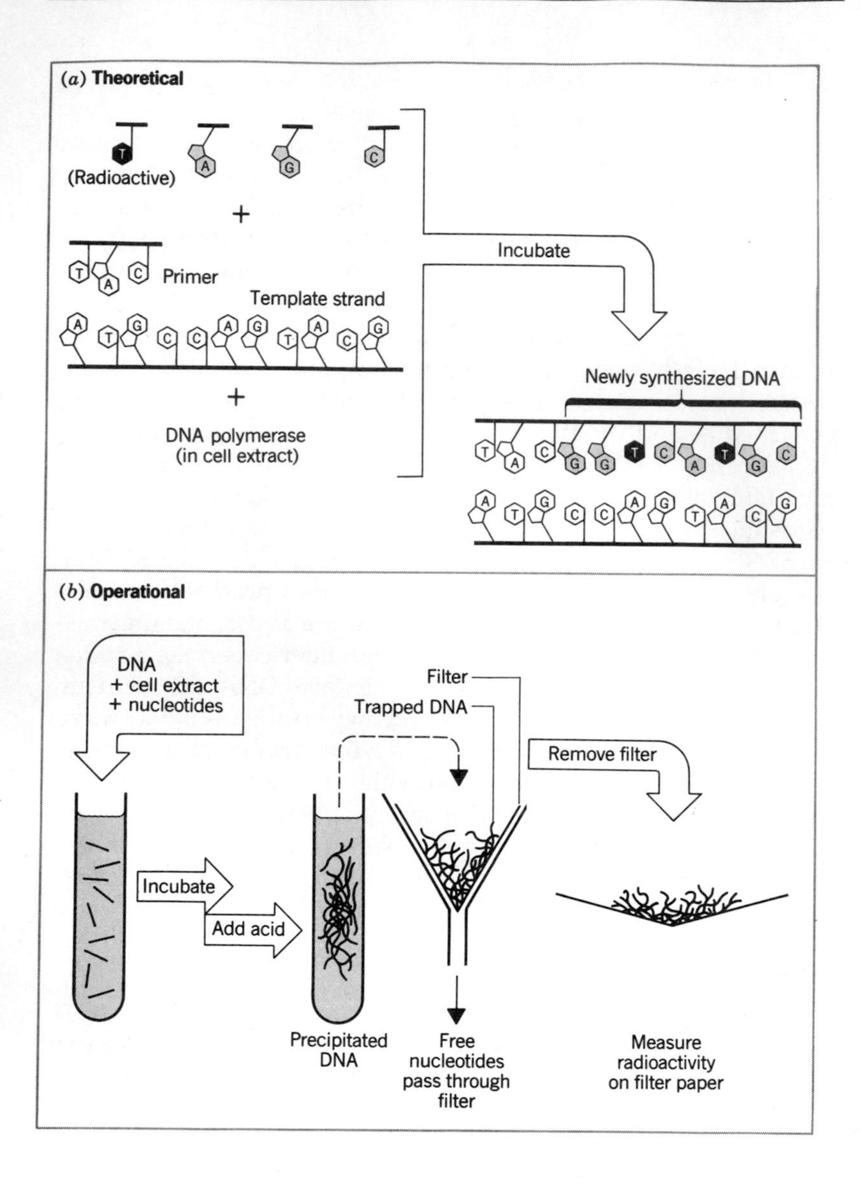
(a) Theoretical
(Radioactive)
T
A
G
C
+
Primer
Template strand
Incubate
+
DNA polymerase
(in cell extract)
Newly synthesized DNA
(b) Operational
DNA
+ cell extract
+ nucleotides
Filter
Trapped DNA
Remove filter
Incubate
Add acid
Precipitated DNA
Free nucleotides pass through filter
Measure radioactivity on filter paper

Molecular biologists use many different enzymes. For each one, a different assay must be devised to follow the enzyme during purification. However, the principle is the same for each assay: there must be a way to distinguish between the reaction **substrate** (the molecules you start with) and the reaction **product** (the molecules you end up with). The molecular biologist measures the speed at which substrate is converted into product. In the example given above, free nucleotides are the substrate of DNA polymerase, and nucleotides incorporated into DNA are the product of the reaction. The speed of the reaction is determined by measuring the amount of radioactivity sticking to the filter paper per minute of incubation time at 37°C for the total reaction mixture. The speed of the reaction indicates the amount of enzyme present in the cell extract (enzymes speed up chemical reactions). Thus, after separating cellular components into different test tubes (Figure 5-7), the molecular biologist uses the assay (Figure 5-8) to determine which tube contains the enzyme.

A number of methods are used to separate subcellular components. In the procedure called **column chromatography**, a glass tube is mounted vertically and filled with one of many different solid materials (e.g., chemically modified **cellulose** powder). The cell extract is allowed to flow slowly through the cellulose packed in the glass tube, eventually dripping out the bottom of the tube. Some molecules bind

---

Figure 5-8 Assay for DNA Polymerase. (**a**) Theoretical. A mixture is prepared that contains the four nucleotides [A, T (radioactive), C, and G], a single-stranded DNA template containing a short, double-stranded region that provides a primer (DNA polymerase always requires a primer to begin its action), and a sample containing DNA polymerase. During incubation DNA polymerase joins the free nucleotides together to form DNA. The newly made DNA will be radioactive because of incorporation of radioactive T. (**b**) Operational. The mixture is added to a test tube and incubated to allow the reaction to occur. Acid is added to stop the reaction and to cause the DNA molecules to clump together. The mixture is poured through a piece of filter paper, and the clumped DNA is trapped on the paper. If any radioactive DNA has been made in the reaction, it will stick to the filter paper because of its large size. In contrast, molecules of T (radioactive) that were not incorporated into DNA are washed through the filter because they are small. The radioactivity on the filter paper is measured with an instrument called a liquid scintillation counter.

tightly to the cellulose, others loosely. Thus, when the biologist passes a dilute salt solution through the cellulose, some molecules come out after very little salt solution has passed through, whereas others require extensive washing. By collecting the salt solution in a series of test tubes (Figure 5-9), the biologist can separate out various mole-

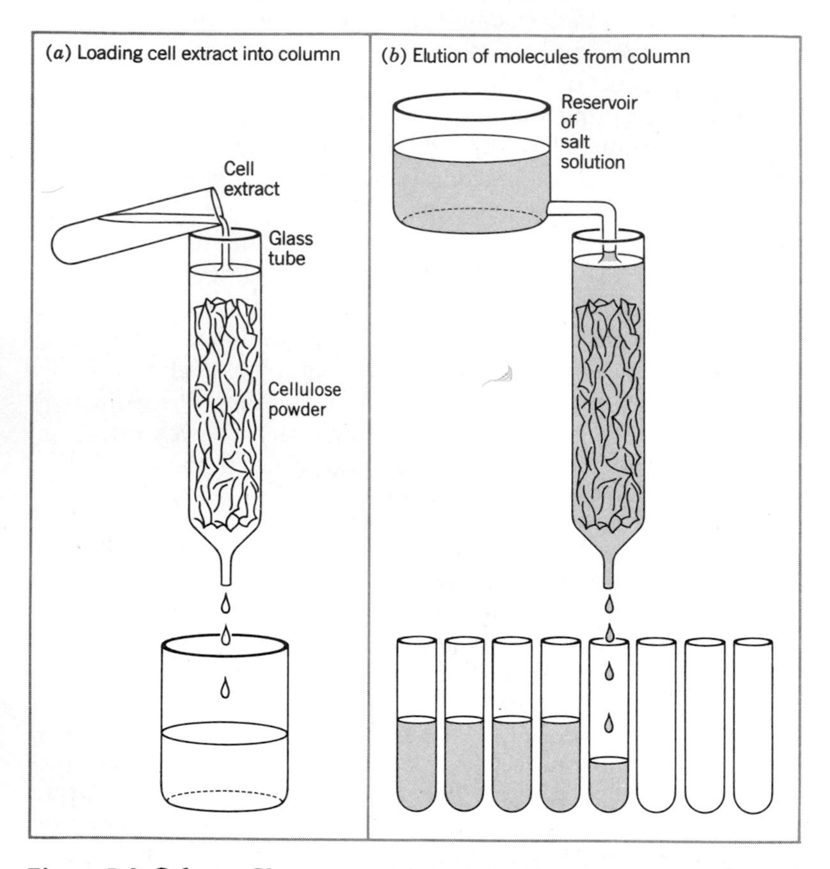

Figure 5-9 Column Chromatography. (a) A cell extract is passed through a solid matrix (often a chemically treated material similar to finely ground cellulose) to which protein molecules stick with varying degrees of tightness. **(b)** The proteins can be removed from the column by passing a dilute salt solution through the column at gradually increasing concentrations. Some molecules will come through sooner than others. All of the salt solution is collected in a series of test tubes, and the contents of the tubes are assayed to determine which tube contains the protein being sought.

cules. Then the contents of each tube can be assayed to determine which one contains the enzyme being sought. By combining several chromatographic procedures, such as the one described above, it is possible to separate an enzyme from all other cellular components.

PROBES TO FIND CLONED GENES

The concepts developed in the discussion of DNA replication can now be used to add important details to the general strategy for cloning genes. DNA is first isolated from cells of interest. It is then cut in specific places with a purified enzyme called a **restriction endonuclease**. This type of enzyme, which serves as the gene cloner's scissors, will be discussed in more detail in Chapter 7. Many DNA pieces are produced, and they are inserted into cloning vehicles using another purified enzyme, DNA ligase. Both enzymes have been purified by strategies similar to those described in the preceding sections, and both are commercially available. The cloning vehicles carry the DNA fragments into *E. coli* cells. The cells are spread onto an agar plate; this physically separates the DNA fragments because different fragments are carried by different cells. The cells grow into colonies, and then a piece of filter paper is placed on the agar plate. Some of the bacteria in each colony stick to this paper; consequently, when the paper is lifted off, some of the bacteria are also removed. The specific pattern of colonies on the plate is preserved on the paper. Next the bacterial cells on the paper are broken in such a way that the DNA sticks to the paper. During this process the double-stranded DNA is converted into single-stranded molecules, thus making the bases available for base-pairing. Then the broken *E. coli* cells are treated with a radioactive **probe** (DNA or RNA) known to be complementary to the gene one wishes to isolate. The probe will form base pairs only with DNA from the particular bacterial cells that contain the gene being sought. This process of forming base pairs between two different **nucleic acid** molecules is called **nucleic acid hybridization** (Figure 5-10). Thus if bacterial cells contain a cloned gene, the DNA associated with those cells will become radioactive when hybridized to the radioactive probe. Clusters of broken, radioactive cells on the filter paper can be identified by placing the paper next to X-ray film, for the radioactive emissions will expose the film. Since the patches of cells on the paper will be in the same pattern as the colonies on the agar plate, the location of

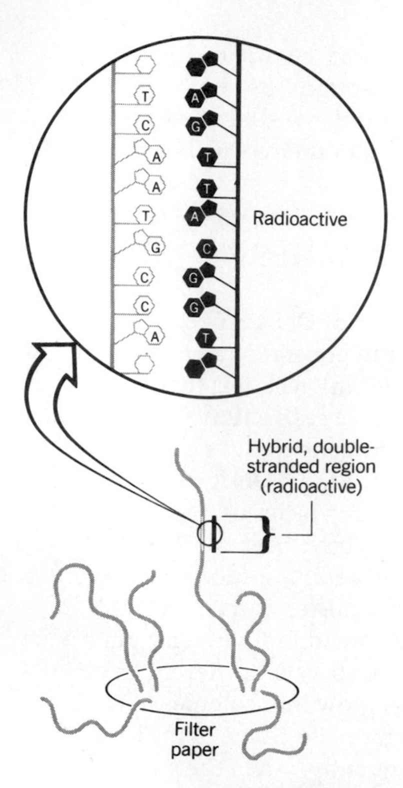

Figure 5-10 Nucleic Acid Hybridization. Under the appropriate conditions two complementary single-stranded nucleic acids will spontaneously form base pairs and become double-stranded. If single-stranded, nonradioactive DNA (shaded) is fixed tightly to a filter and then incubated in a solution containing single-stranded, radioactive DNA (solid), double-stranded regions will form in which the two types of DNA have complementary nucleotide sequences; the radioactive DNA will become indirectly bound to the filter through its attachment to a specific region of nonradioactive DNA (open). By measuring the amount of radioactivity bound to the filter, one can estimate the relatedness of two DNAs.

the radioactivity on the paper indicates the colonies on the agar plate that contain the cloned gene.

Radioactive probes can be obtained in several ways. In some cases it is possible to isolate messenger RNA for the gene one wishes to clone. Radioactive messenger RNA to be used as a probe can be obtained by growing the cells in the presence of radioactive nucleotides. However, it is often difficult to obtain natural messenger RNA containing enough radioactivity to test the bacterial colonies for cloned genes because so much radioactivity must be added to the cells that they tend to die. Consequently, the messenger RNA is usually employed as a template for the synthesis of DNA using **reverse transcriptase**, an enzyme purified from **RNA tumor viruses**. Since the DNA product, called **complementary DNA** (cDNA), is formed in test tubes, it can be made highly radioactive by using radioactive nucleotides to form the DNA.

A more general way to obtain a probe involves first purifying the protein product of the gene being sought. The order of the amino acids in the protein can then be determined by chemical analysis. Since every amino acid corresponds to a triplet of nucleotides in the DNA of the gene, it is possible to predict the nucleotide sequence of the gene from the sequence of amino acids (the prediction is not exact because most amino acids are encoded by several different codons; see Figure 3-6). The next step is to chemically synthesize a short stretch of DNA, one base at a time, that would have a sequence identical to that predicted for a region of the gene. Instruments are available to determine the amino acid sequence of the protein and to synthesize the short DNA piece. A purified enzyme called a **kinase** is used to add radioactive phosphorus to the synthetic DNA fragment, making it a highly radioactive probe.

DNA AMPLIFICATION

Principles derived from studies of DNA replication have also been used to selectively synthesize short regions of DNA, sometimes cutting months from gene cloning procedures. In the process called **polymerase chain reaction** a specific segment of DNA can be enriched by more than a hundred thousand times relative to nearby nucleotide sequences. This process is outlined in Figure 5-11. A key concept is that DNA polymerase always requires a primer to begin DNA synthesis. A primer is the end of a preexisting segment of either DNA or RNA to which DNA polymerase adds nucleotides. Thus DNA polymerase cannot begin synthesis anywhere along a single-stranded DNA molecule but must start at points defined by primers. If the two strands of a long, double-stranded DNA molecule are separated and a short

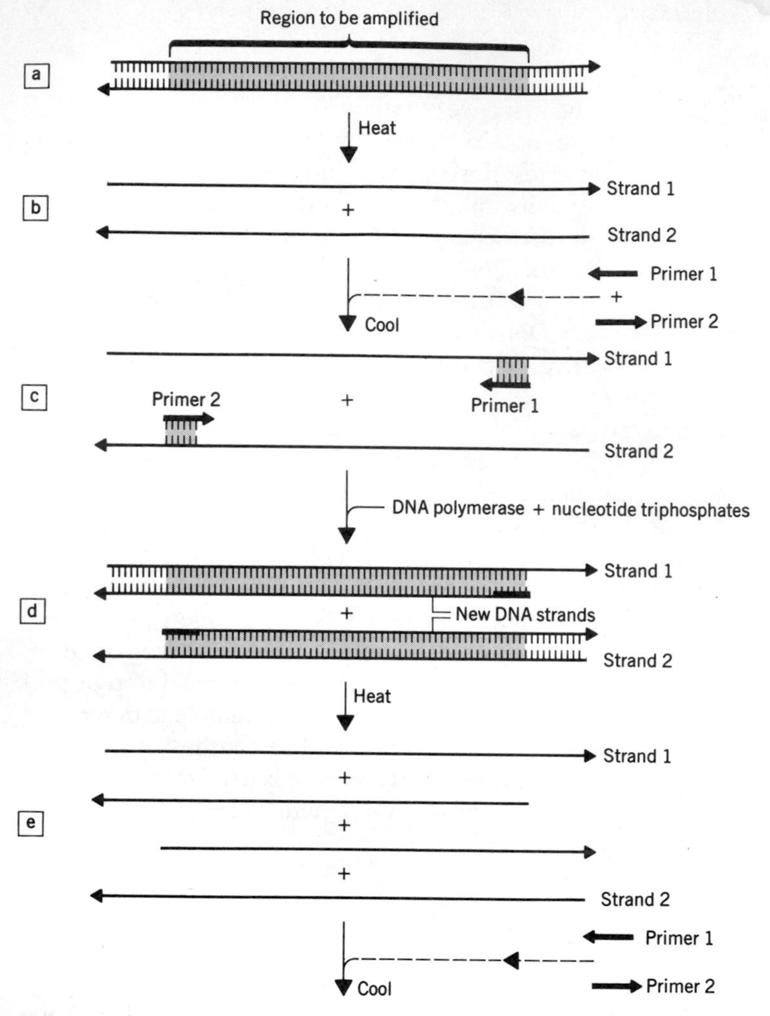

Figure 5-11 Amplification of DNA. (**a**) A double-stranded DNA molecule that contains a specific region of interest (double-strandedness is indicated by vertical lines). (**b**) Heating the DNA causes the two strands to separate. (**c**) When primers complementary to short regions on each of the two strands shown in (a) are added to the mixture and it is cooled, the primers hybridize to the two strands labeled **1** and **2**. (**d**) The primers are extended by the addition of DNA polymerase and nucleotide triphosphates. The new DNA molecules have different lengths because of the asynchrony of the polymerization reaction. (**e**) If the mixture is heated, all of the DNA will become single-stranded. (**f**) Upon cooling, primers will hybridize to both old and new strands in the

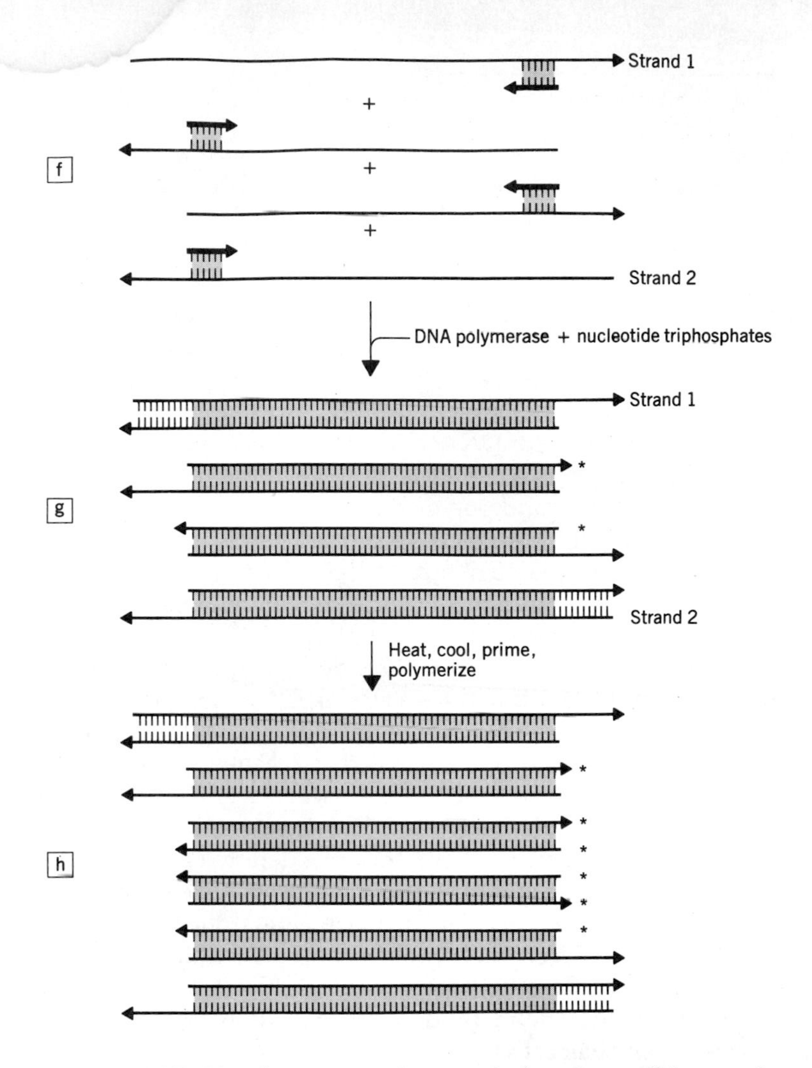

mixture. (**g**) DNA polymerase again extends the primers. If the templates are the strands made in step **d**, DNA synthesis stops when the polymerase reaches the end of the template. This produces a discrete DNA fragment (*) containing the nucleotide sequences of both primers and the DNA in between. (**h**) A subsequent cycle of heating, cooling, and polymerization increases the relative abundance of the discrete fragment (*). Heat-resistant DNA polymerase is used, so if enough polymerase, nucleotide triphosphates, and primers are added at the beginning of the reaction, the process consists only of heating and cooling steps.

primer is hybridized to one strand at a specific spot (Figure 5-11*c*), DNA synthesis will always begin there. Because of the directionality of polymerase movement along DNA, synthesis will proceed in only one direction from the primer. A different primer can be placed on the second strand to permit the synthesis of the region between the primers (see Figure 5-11*d*). After DNA synthesis has occurred, the mixture is heated to produce single-stranded molecules from the double-stranded ones. Upon cooling, the primers will hybridize to the new DNA as well as to the original strands. Another round of DNA synthesis generates discrete DNA fragments, which include the sequences of both primers and the DNA in between (Figure 5-11*g*). The cycle of heating, cooling, and polymerization is repeated many times, and with each cycle the discrete fragment increases in abundance. Within 3 hours it is possible to obtain a specific fragment of DNA that can be easily purified. The limiting factor is knowing enough about the nucleotide sequence to generate the primers.

One major application of gene amplification is in genetic screening. The amplification process generates a specific DNA fragment whose nucleotide sequence can be easily determined, making it possible to detect the presence of mutations associated with genetic diseases. The method can also be used to detect viral diseases such as AIDS. The nucleotide sequence of the viral genome is known, and if it is present in even one in a thousand human cells, it can be detected by amplification. A third application concerns the identification of people. Although all humans have very similar nucleotide sequences in their DNA, each person's DNA is slightly different. Thus amplification and nucleotide sequence determination of particularly variable genes can replace fingerprinting. The method is so sensitive that blood stains, skin fragments, and hair cells may provide enough material for analysis. Amplification will become increasingly important in criminal cases, for it will greatly increase the ability of law enforcement agencies to connect particular individuals with a crime.

PERSPECTIVE

By the early 1950s the importance of DNA to heredity was widely recognized, and when Watson and Crick discovered how the two DNA strands are arranged, biochemists set out to determine how DNA replication works. Enzyme assays and protein purification procedures were developed, and by the early 1970s a number of proteins that participate in DNA replication had been purified. Consequently, when methods were found to cut DNA into specific pieces, replication proteins that would join the pieces were already available. In addition, other replication proteins were known that could synthesize highly radioactive DNA to be used as probes for locating bacterial colonies containing cloned genes. Thus information from a number of lines of research was used to develop gene cloning strategies. These strategies are now being used to study aspects of DNA replication that we still don't fully understand, such as how chromosome replication begins and ends.

Questions for Discussion

1. Why is complementary base pairing important for understanding how DNA is duplicated?
2. During the normal cycle of DNA replication, each of the two parental DNA strands is replicated; the result is two new double-stranded DNA molecules which are identical. But some viruses (see Chapter 6) contain only a single strand of DNA. Others contain only a single strand of RNA. How might their genetic material be replicated?
3. The circular chromosome of *E. coli* is replicated in about 40 minutes by two forks proceeding in opposite directions from an origin of replication to a termination point 180 degrees from the origin. The chromosome is about 4 million base pairs long. If there are 10 base pairs per turn, calculate how fast the DNA must rotate along its long axis, in revolutions per minute, to unwind as the forks move.
4. If it takes *E. coli* 40 minutes to replicate its chromosome, how can this bacterium divide every 20 minutes with each daughter cell getting a complete copy of the chromosome?
5. When double-stranded DNA molecules are heated sufficiently, the two strands separate. This process is called denaturation or melting, and often it occurs within a narrow temperature range

(on the order of 5 °C). The temperature at which melting occurs depends on the strength of the forces holding the strands together. If you form hybrids between two nucleic acids that are not perfectly complementary, would the melting temperature be higher or lower than that of a perfect hybrid? Why?

6. Many chemicals in our environment cause mutations in bacteria. Describe a way in which mutation frequency could be used to test new chemicals before they reach the environment.
7. Occasionally a point mutation will occur in a gene, rendering its protein product nonfunctional (see Figure 5-5). Cases have been found in which a second mutation, located in another gene, causes the cell to produce a normal, functional protein. This second mutation is called a suppressor mutation. Often suppressor mutations fall in the anticodon region of transfer RNA genes. Using Figure 3-6, determine which nucleotide change or changes in the anticodon of the tRNA would suppress the mutation illustrated in Figure 5-5. Would you expect such a tRNA-type suppressor mutation to affect mutations in many genes, or would it be gene-specific?

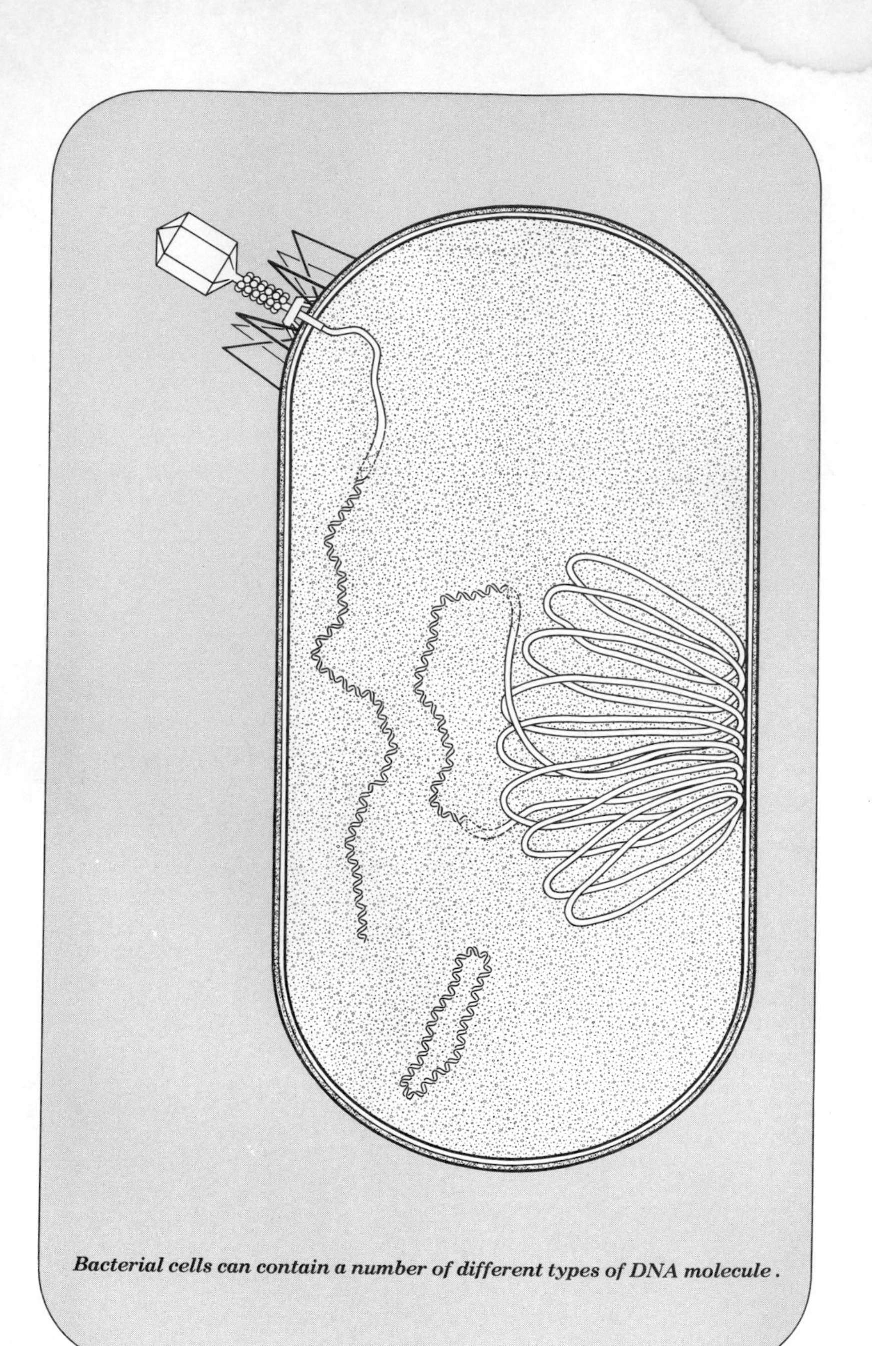

Bacterial cells can contain a number of different types of DNA molecule .

CHAPTER SIX

PLASMIDS AND PHAGES

Submicroscopic Parasites as Cloning Tools

Overview

Bacterial plasmids and phages are submicroscopic agents that infect bacterial cells and then use the bacterial components to replicate themselves. They contain genetic information, which in most cases is stored in short DNA molecules. Molecular biologists have found plasmid and phage DNA molecules particularly easy to handle and study. Genetic information from animals and other organisms having very long DNA molecules is difficult to study unless the DNA has been cut into small pieces and the pieces have been physically separated. Plasmids and phages assist us in separating DNA pieces. Since DNA fragments can be inserted into plasmid or phage DNA without impairing infectivity, these tiny infectious agents can be used as vehicles to place almost any DNA fragment inside a living bacterial cell. There the DNA fragment reproduces as a part of the plasmid or phage DNA.

INTRODUCTION

Certain types of small DNA molecule are infectious. Once inside a living cell, they can utilize the machinery of the cell to reproduce. Infectious DNAs can have profound effects on living cells. Some types take over a cell and kill it, while other types can be beneficial to the host. These infectious DNA molecules fall into two general types, the **plasmids** and the **viruses**. Plasmids are naked DNA molecules; they are generally circular and are found inside cells only. Viruses surround their DNA molecules with a protective shell of protein; thus, they can sometimes survive for many years outside their host cell. A virus that infects a bacterium is called a **bacteriophage** or simply a **phage**.

It is possible to cut plasmid and bacteriophage DNA in a specific place, insert a piece of DNA from a different source, and still retain all the information necessary for infection and replication by the plasmid or phage. Thus these infectious DNA molecules are useful as tools to transfer DNA from one type of cell to another; they are the cloning vehicles referred to in earlier chapters.

Before delving into what happens when DNA molecules invade cells, it is useful to restate the problem that cloning vehicles help overcome: the specific fragment of DNA one wishes to obtain must be separated from the thousands (or sometimes millions) of other fragments produced during the cutting and joining process, and then the fragment must be located. The separation aspect is not a problem per se; one could place a drop of water containing DNA fragments on an agar plate, smear the drop over the whole surface of the plate, and thus easily separate the fragments. But then it would be very difficult to detect the fragments; they are so small that they are visible only by using an electron microscope, and this instrument is impractical for scanning a large surface or for distinguishing one stretch of DNA from another. Moreover, DNA fragments spread on an agar plate would be too dilute to find with a complementary radioactive probe (described in Chapters 5 and 8). In one sense, such a spreading process would be much like scattering straw from a haystack over a large field. Finding one particular straw would be difficult. The problem could be solved, however, if the straw of interest were known to have seeds attached. When the seeds sprouted into a plant, the plant could be seen and the particular straw would be found at the base of the plant.

In the case of DNA fragments, one needs a way to multiply the fragments after scattering them; then radioactive probes can be used to find specific ones. Gene cloners multiply the fragments using cloning vehicles and bacterial cells. First, the vehicles and their attached DNA fragments are transferred into bacterial cells (one vehicle and fragment per cell). Next, the cells are scattered on the surface of an agar plate. Third, the cells are allowed to multiply. The cloning vehicles, along with the attached DNA fragments, also multiply.

It might appear that cloning vehicles are unnecessary in the process described above. One need only get the DNA fragments into bacterial cells so that each cell obtains but a single fragment; then the cells can be spread out on an agar surface. Each cell will multiply to form a colony, and all the colonies can be tested for the gene of interest. Indeed, many types of bacteria will ingest DNA through their cell walls, but very few DNA fragments have the necessary start and stop signals to cause the cellular machinery in bacteria to replicate the DNA fragment. If no DNA replication occurs, the fragment will be diluted out as the bacteria grow and divide, for only the original bacterial cell will contain the DNA fragment. Even after many cell divisions, only one cell in the colony will contain the fragment. To bypass this problem, DNA fragments are first inserted into plasmid or phage DNA because they contain the correct signals for replication. The plasmid and phage DNAs used as cloning vehicles then enter the bacterial cells and multiply as the cells multiply.

PLASMIDS

Most bacterial plasmids studied to date are small, circular, double-stranded DNA molecules (Figure 6-1) that occur naturally in their hosts. Like all natural DNA molecules, plasmids contain a special re-

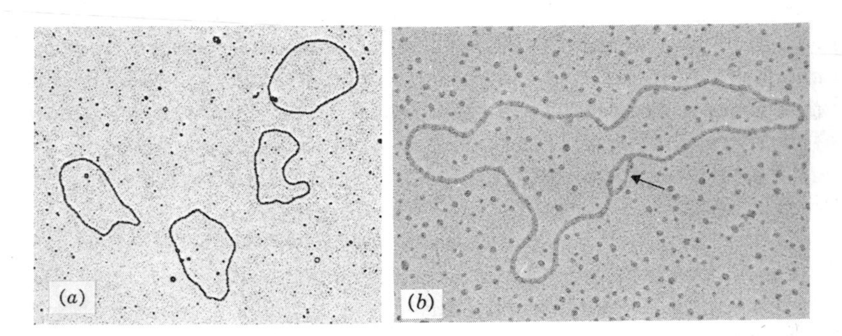

Figure 6-1 Electron Micrographs of Plasmids. (**a**) Four DNA molecules of the type called ColEl. These small, circular DNAs are only 0.001 times the length of *E. coli* DNA (compare with Figure 1-1). (Photomicrograph courtesy of Grace Wever, Eastman Kodak) (**b**) Enlargement of a plasmid similar to ColEl. Sample preparation conditions were adjusted so that a short region of DNA would become single-stranded (arrow). (Photomicrograph courtesy of G. Glikin, G. Gargiulo, L. Rena-Descalzi, and A. Worcel, University of Rochester.)

gion in their DNA called an **origin of replication**. The origin serves as a start signal for DNA polymerase and ensures that the plasmid DNA molecule will be replicated by the host cell. Plasmids differ in length and in the genes contained in their DNA. Some of the smaller plasmids, which are popular in gene cloning, have about 5000 nucleotide pairs, enough DNA to code for about five average-sized proteins. In comparison, *E. coli* contains slightly more than 4 million nucleotide pairs in its DNA, and we have about 4 billion nucleotide pairs in ours. Many of the larger plasmids are difficult to handle and are transmissible from one bacterial cell to another. Thus they are generally not used in gene cloning; however a brief description of one type is included in Appendix I for interested readers.

An important aspect of plasmid DNA molecules is that they often make their host bacterial cell resistant to **antibiotics**. Antibiotics generally block processes vital for cell function. For example, **penicillin** interferes with the formation of bacterial cell walls. Some plasmids carry a gene for an enzyme that destroys penicillin. Therefore, bacteria containing those plasmids are able to grow in the presence of penicillin and are termed resistant.

Drug resistance turns out to be extremely useful in genetic engineering. For example, in the cloning procedure outlined earlier, DNA fragments are inserted into a plasmid DNA having a gene that confers resistance to penicillin. Then the plasmid DNA is added to a culture of bacteria that normally are killed by penicillin. Under the proper experimental conditions, the plasmid DNA enters the cell by a process called **transformation**. It then multiplies along with the bacterial cell. The bacteria are next spread on an agar plate containing penicillin and incubated overnight. Most of the bacteria are killed because few cells actually acquire a plasmid during transformation. However, the few cells that do obtain a plasmid are penicillin resistant, and they form colonies. Every colony growing on the agar plate contains cells harboring copies of the plasmid. Thus, when testing colonies for specific genes, gene cloners use antibiotics to avoid examining the millions of bacterial colonies that *fail* to be transformed by a plasmid.

OBTAINING PLASMID DNA

After a gene cloner has inserted DNA fragments into plasmids, transferred the plasmids into bacterial cells, selected for cells transformed to antibiotic resistance by the plasmid, and determined which colony contains the desired DNA (Chapter 8), the cloned DNA fragment must be retrieved by purifying the plasmid carrying it.

The first step in obtaining plasmid DNA is to prepare a liquid bacterial culture containing billions of cells harboring plasmids (see Chapter 2 for culture techniques). Each of these bacterial cells contains many copies of the circular plasmid carrying the desired DNA fragment.

Next the DNA molecules must be removed from the cells. Bacterial cells that have grown in broth culture are first concentrated; one procedure is to place the broth in a test tube and allow the cells to gravitate to the bottom of the tube much like silt settling out of river water. Generally molecular biologists speed up the settling process by putting the test tube in a centrifuge as described in Figure 5-7. The bacteria are driven to the bottom of the tube, where they form a tight pellet. The bacteria stay as a pellet even after the tube has been removed from the centrifuge, so the broth they were growing in is easily poured off. A small volume of water is then added to the trillion or so bacteria, and the tube is shaken to distribute the bacteria in the water. Enzymes and detergents are added to this suspension to dissolve the cell walls of the bacteria, releasing both bacterial DNA and plasmid DNA molecules from the cells. At this stage the content of the test tube is called a **cell lysate**.

Once the DNA molecules have been freed from the cells, the plasmid DNA must be physically separated from the bacterial DNA. The two types of DNA differ mainly in their length. Depending on the particular plasmid, the bacterial DNA may be up to a thousand times longer. Since bacterial DNA is so large, it tends to sediment to the bottom of the test tube faster than the small plasmid DNA. Consequently, one step in the purification procedure is to put the cell lysate in a tube that is spun in a centrifuge. The large bacterial DNA will form a pellet in the bottom of a test tube while the much smaller plasmid DNA will stay in the upper fluid. Unfortunately, there is usually so much more bacterial DNA than plasmid DNA that this centrifugation procedure does not completely separate the two kinds of DNA molecule.

The great length of the bacterial DNA, however, makes it possible to carry out an additional type of separation. Bacterial DNA is easily broken by sucking the DNA-containing solution into a pipette. This procedure does not break the smaller plasmid DNA; after a few squirts through the pipette, bacterial DNA becomes linear while the plasmid DNA remains circular. Circular DNA has a distinctive property that allows it to be separated from linear DNA. This property is related to the concepts of buoyancy and relative density.

Consider for a moment your own buoyancy in water. If you are totally relaxed and motionless, you tend to float so that only the top of your head is above the surface. You can change your buoyancy in two ways: by putting on a life jacket or by holding some rocks in your hands. The life jacket lowers your overall buoyant density, causing you to float higher. The rocks increase it, causing you to sink lower. In addition to your own density, the density of the water is important in determining whether you sink or float. For example, water containing a high concentration of salt has a high density, and as a result people easily float in very salty water such as that of the Dead Sea.

In the laboratory, test tubes can be filled with salt solutions so that the salt concentration gradually increases from top to bottom. The gradual change in salt concentration is called a concentration gradient, and it produces a **density gradient**. Salt concentrations can be adjusted so that molecules such as DNA will sink until they reach a solution density equal to their own density. At that point the DNA is at **equilibrium**, and if unperturbed, it will remain at that position forever. The important point to remember is that the depth to which a DNA molecule sinks in a density gradient depends on the density of the DNA and the density of the solution.

Now let's return to circular and linear DNA molecules. Both have the same four base pairs, so physically they should have the same density. However, it is possible to add a dye molecule to DNA that acts as a life jacket, lowering the density of the DNA. Linear DNA molecules can bind more dye than circular ones; linear molecules can effectively put on more life jackets, hence will float higher. Why do more dye molecules bind to linear DNA than to circular DNA? As the dye molecules bind to DNA, they insert themselves between the base pairs and slightly unwind the DNA (Figure 6-2*a*). Unwinding causes the DNA to twist. Extreme examples of unwinding are shown in Figure 6-2*b* and *c* to illustrate the difference between linear and circular DNA. Twists introduced into a linear DNA molecule are quickly lost as the free ends of the molecule rotate over each other (Figure 6-2*b*). In contrast, twists put into a circular molecule are retained because there are no free ends (Figure 6-2*c*). As more and more dye molecules bind to circular DNA, the twists accumulate, and it becomes increasingly difficult for subsequent dye molecules to bind. This is not the case with linear DNA. Thus the absence of ends for strand rotation prevents a circular DNA molecule from binding as much dye as a linear one.

Operationally the process is much simpler than the explanation. A DNA preparation is mixed with the dye and a heavy salt in a test tube. At this stage one could rely on the force of gravity to generate the density gradient with the salt water, but it would take a very long time for the salt molecules to accumulate in the bottom of the tube. It would also take a long time for the DNA molecules to settle to their own density. The time required can be shortened by putting the tube containing DNA, salt, and dye in a centrifuge. After spinning the sample, the tube is examined with ultraviolet light (black light). When dye molecules bound to DNA absorb the ultraviolet light, they emit a visible, fluorescent light. Two bands can be observed in the tube. The upper one corresponds to the linear DNA (bacterial DNA) and the lower one to the circular plasmid DNA (see Figure 6-3). The circular plasmid DNA can be sucked out of the tube with a pipette, which will then contain pure plasmid DNA.

Following plasmid isolation, the cloned genes can be cut out of the plasmid DNA by restriction endonucleases as described in Chapter 7. This produces a small number of DNA fragments; the fragments can be separated from each other by gel electrophoresis (see Figure 7-4), the resulting DNA bands can be removed from the rest of the gel by cutting with a razor blade, and the DNA can be washed out of each small gel piece.

BACTERIOPHAGES

Plasmids are naked DNA molecules containing an origin of replication and at least one gene encoding a protein involved in regulating DNA replication. Bacteriophages, commonly called phages, are viruses that infect bacteria, and they are more complicated than plasmids. In addition to having an origin of replication, phage DNA contains genes coding for proteins that form a protective coat around the DNA. But, like plasmids, phages lack the machinery necessary to actually make proteins; consequently, they reproduce only inside living bacterial cells. Both phages and plasmids can be used to separate and amplify specific DNA fragments, but the two have very different means of re-

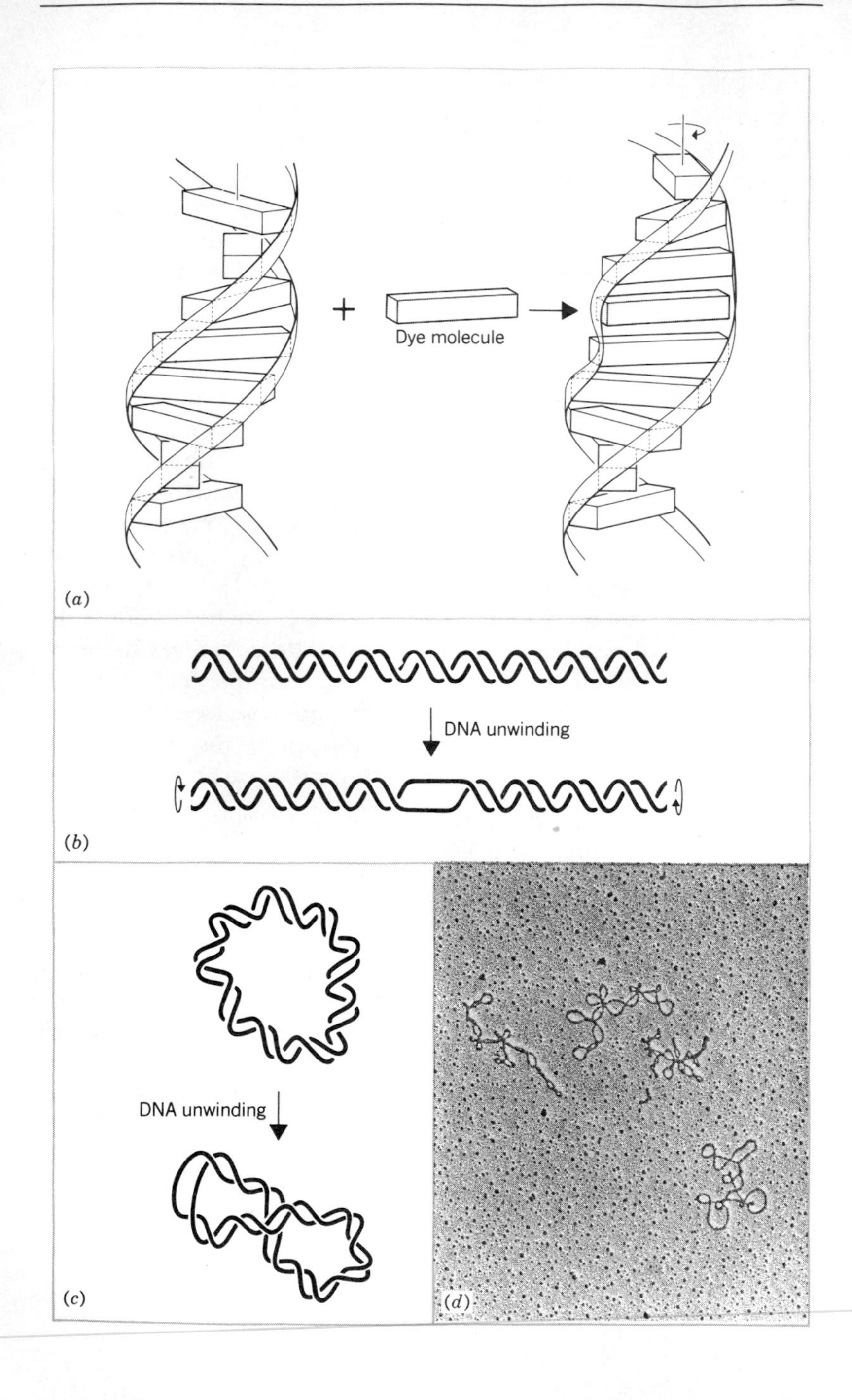

(a)
(b)
(c)
(d)

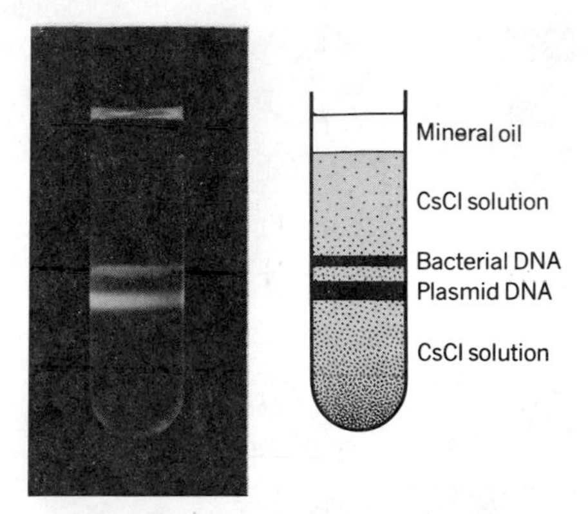

Figure 6-3 Separation of Plasmid and Bacterial DNAs by Dye Bouyant Density Centrifugation. Plasmid DNA, bacterial DNA, water, dye (ethidium bromide), and a heavy salt (cesium chloride: CsCl) were mixed in a plastic tube and centrifuged for two days at 35,000 rpm. Before centrifugation, mineral oil was added to fill the plastic tube to prevent its collapse from the force of the centrifugal field. After centrifugation, the tube was illuminated with ultraviolet light, and bright orange bands appeared in the tube, indicating the location of the DNA molecules.

production. Thus phages and plasmids are used for different aspects of cloning procedures.

Some phages are like miniature hypodermic syringes (Figure 6-4). The phage DNA is wrapped into a tight ball inside a headlike structure made of proteins. A tail, also made of proteins, is attached to the head.

Figure 6-2 Binding of Dye to Linear and Circular DNA Molecules. (a) Dyes such as ethidium bromide are flat structures that resemble DNA base pairs. When such a dye binds to DNA, it slips in between two adjacent base pairs (shown as blocks). Although the hydrogen bonds holding the base pairs are not broken by the dye, the DNA double helix unwinds slightly (26 degrees per dye molecule bound). **(b)** Unwinding twists the DNA, but with linear DNA the twisting dissipates as the ends of the DNA rotate. For illustrative purposes, an extreme case of unwinding is shown, and base pairing has been disrupted. **(c)** No free ends are present in circular DNA, so twists arising from unwinding accumulate. The twists make further unwinding, and thus dye binding, more difficult. **(d)** Electron micrograph of twisted plasmid DNA molecules. (Photomicrograph courtesy of G. Glikin, G. Gargiulo, L. Rena-Descalzi, and A. Worcel, University of Rochester.)

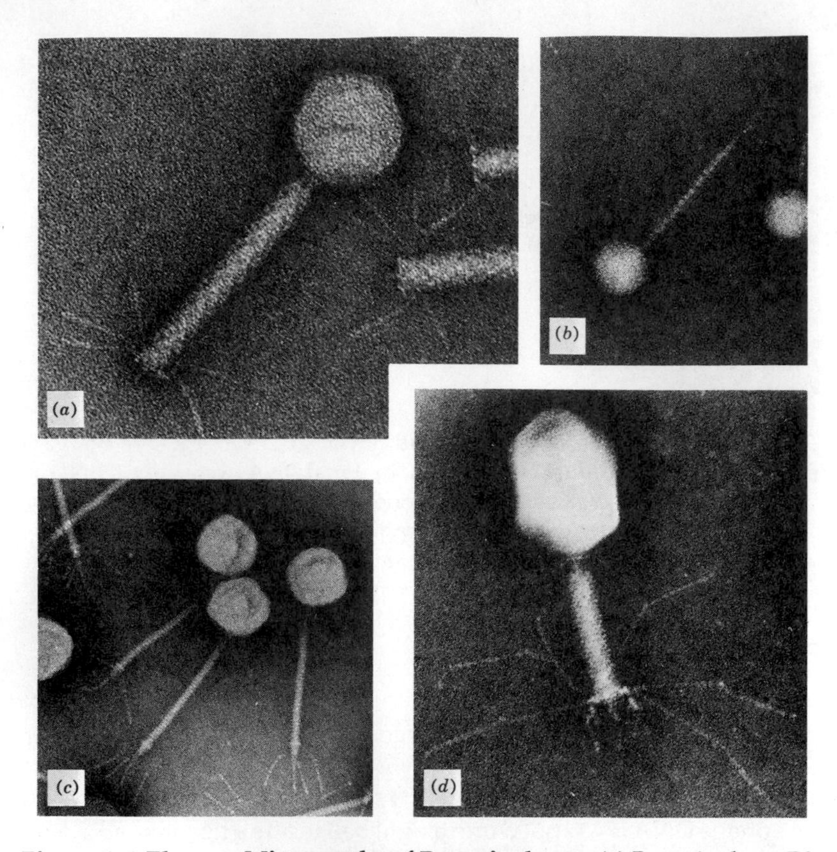

Figure 6-4 Electron Micrographs of Bacteriophages. (**a**) Bacteriophage P2, magnification 226,000 times. (**b**) Bacteriophage lambda, magnification 109,000 times. (**c**) Bacteriophage T5, magnification 91,000 times. (**d**) Bacteriophage T4, magnification 180,000 times. (Photomicrographs courtesy of Robley Williams, University of California, Berkeley.)

When such a phage particle comes in contact with a bacterial cell, the phage tail sticks to the cell wall, and the DNA is squirted out of the head, through the tail, and into the bacterium (see frontispiece, Chapter 6). Soon after the phage DNA gets into the cell, it begins to take control. Special phage genes are transcribed by the bacterial RNA polymerase, and the resulting messenger RNAs are translated into phage proteins using the bacterial ribosomes. At early stages of infection some phages produce proteins that destroy the bacterial DNA, chopping it into individual nucleotides. Once that has happened, the bacterium is doomed, because all the information needed for its own reproduction is gone. Some phages have genes that produce an RNA polymerase, so they do not have to rely on the host polymerase to make messenger RNA from phage genes.

Many phages also have genes for their own DNA replication machinery. When this apparatus is in place, the phages can use nucleotides released from the bacterial DNA to make phage DNA. Hundreds of copies of the phage DNA are made, and within minutes other genes on the phage are turned on to produce new head and tail proteins. The head proteins assemble into heads, phage DNA is packaged inside them, and a tail is attached to each head. The assembly of new phages occurs *spontaneously*, and the total time from injection of DNA to production of new phages can be less than 20 minutes. The bacterium becomes little more than a shell containing hundreds of new phage particles. As a final act, the phage produces an enzyme that destroys the bacterial cell wall, releasing the phage particles to seek new hosts.

Phage efficiency is awesome. One phage can produce hundreds of **progeny** particles. Each progeny particle can infect a bacterial cell and produce many more phage particles. By repeating the infection cycle just four times, a single phage particle can lead to the death of more than a billion bacterial cells.

Next, consider what happens if a small number of phage particles is added to a *dense* bacterial culture, and before the first bacterial cell has been broken open by the newly made viruses, the culture is spread on an agar plate. Within 20 to 30 minutes, the first infected bacteria rupture, releasing phage particles. So many bacteria are on the agar plate that the new viruses quickly attach to nearby bacteria and repeat the infection process. Gradually the circle of death expands away from the point at which the original infected cell had fallen on the agar surface. Meanwhile, the uninfected cells, which are the vast majority, continue to divide. They are unaffected by the fact that a few of their number are infected, and as they become more numerous, they begin to deplete the food supply. Eventually, the uninfected bacteria completely cover the agar surface except for the small regions where the phages are attacking the cells. Within a day the bacteria stop growing, and their biochemical machinery shuts down. The phages, too, stop reproducing, since they rely on active bacterial machinery to supply energy for their enzymes. Thus the agar plate is covered by a lawn of bacteria containing small holes wherever the phages had been multiplying

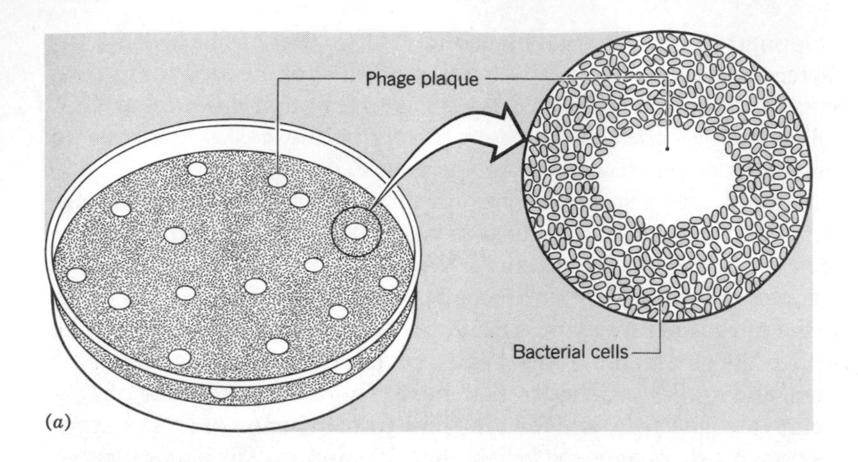

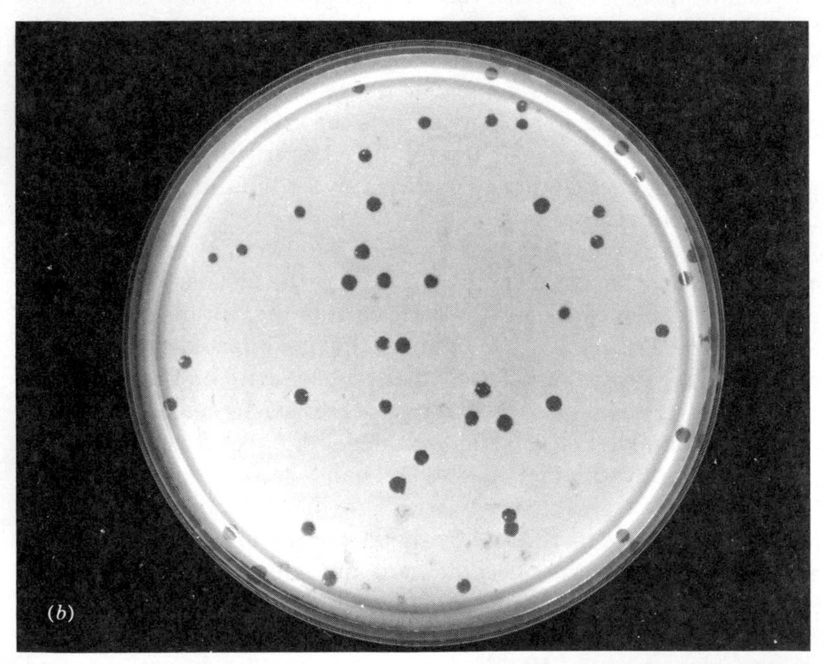

Figure 6-5 Bacteriophage Plaques. An agar plate is covered by a lawn of bacteria. The holes in the lawn, called plaques, are regions where phages have killed the bacteria. (**a**) Schematic diagram. (**b**) Photograph of agar plate. (Photograph courtesy of Robert Rothman, University of Rochester.)

(Figure 6-5). These holes, called **plaques**, are about an eighth of an inch in diameter, and each one arose from a single phage particle.

If a DNA fragment is inserted into a phage DNA molecule without destroying important phage genes, the phage will reproduce the fragment along with its own DNA when it infects a bacterial cell. Gene cloners can determine which plaque has a particular piece of DNA by using a radioactive probe to test the plaques for DNA having base pairs complementary to a specific gene. Once the right plaque has been found, a procedure similar to that described previously for bacterial colonies is used to obtain large amounts of the gene.

First, a piece of sterile wire is poked into the plaque containing the cloned gene. A small number of virus particles will attach to the wire. When the wire is stuck into a fresh culture of bacteria, the phages drop off the wire, attack the bacteria, and reproduce billions of times. Large amounts of phage DNA will be made and packaged. Since the cloned gene is actually a part of the phage DNA, it too will be very abundant; moreover, it will be packaged inside a phage head along with the phage DNA sequences. Phage particles can be easily purified by density gradient centrifugation in a way similar to that described earlier for purification of plasmid DNA. Since DNA and protein have different densities, phages, which are a combination of DNA and protein, will have a buoyant density between that of pure DNA and pure protein. Thus, no dyes are needed to separate phages from other cellular components, including bacterial DNA. A tube containing the phage in a heavy salt solution is centrifuged until a density gradient is established and the phage sediments to its own density. When the tube is removed from the centrifuge and examined, the phage will appear as an opalescent band that can be easily sucked out with a pipette (Figure 6-6).

One of the phages used for cloning is called **lambda** (Figure 6-4*b*). When lambda DNA is injected into a bacterial cell, it has two choices. It can behave as described above and destroy the bacterium. Alternatively, it can take up residence in the cell and enter a quiescent mode. When the latter choice is made, the lambda DNA inserts (**integrates**) into the bacterial chromosome; it becomes part of the bacterial DNA (Figure 6-7). Our knowledge of integration, a natural joining of DNA molecules, is important for understanding a number of important phenomena including the biology of AIDS (Chapter 11). Accompanying integration is a change in phage gene control. The phage genes that

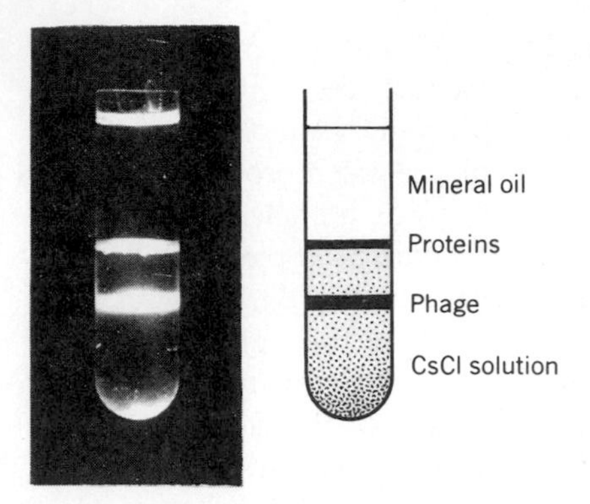

Figure 6-6 Purification of a Bacteriophage by Centrifugation. A bacterial lysate containing phage particles was placed in a plastic tube, mixed with cesium chloride, and centrifuged at 25,000 rpm for one day. During the centrifugation the phage particles formed a band as indicated in the figure. The band above the phage is composed of bacterial proteins and cell wall material. Above this band is a layer of mineral oil used to fill the tube to prevent collapse during centrifugation.

normally would produce the proteins to kill the cell are turned off by a repressor protein made from a phage gene. Thus every time the bacterial DNA replicates, lambda DNA replicates along with it. In this dormant state lambda DNA does little more than produce repressor protein to keep its genes shut down. At the same time, the repressor protects the bacterial cell from infection by other lambda phages—when the newcomers inject their DNA, it is quickly bound by a repressor produced from the resident lambda. All genes necessary for replication are shut off, and thus the incoming phage is unable to initiate a **lytic infection** that would kill the cell. Consequently, one can easily find **lysogens** (bacterial cells that are being protected by a resident phage) by looking for bacterial colonies in the middle of a phage plaque.

If the repressor that shuts down lambda gene expression is destroyed, the phage DNA removes itself from the bacterial chromosome and directs the cell to make phage particles. The bacteria then become filled with phage, break open, and release phage into the environment.

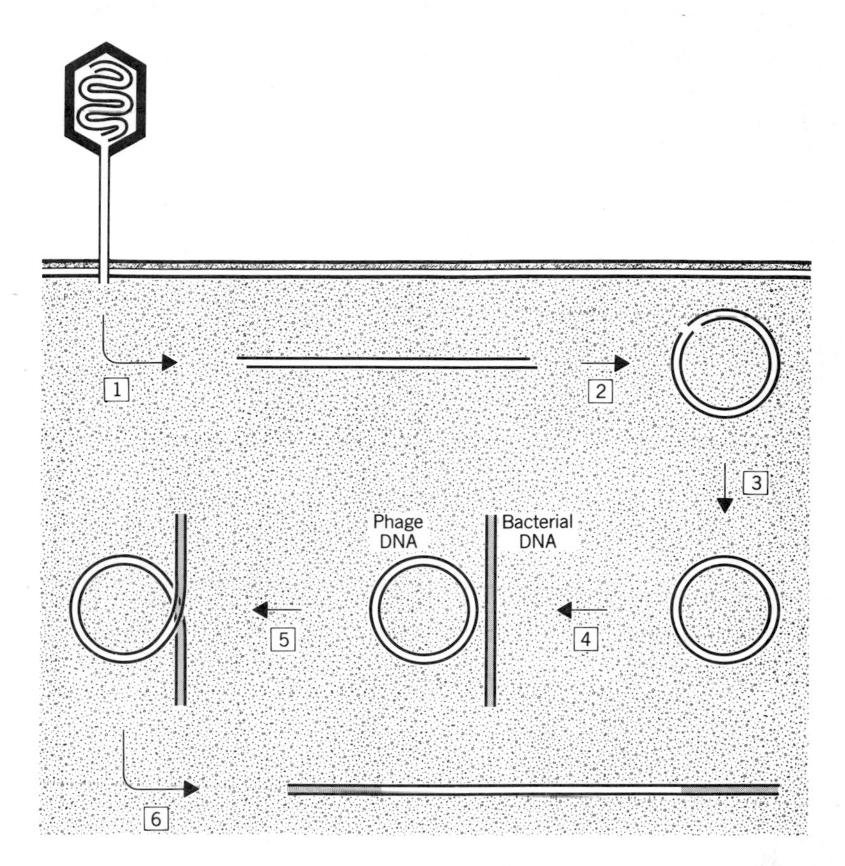

Figure 6-7 Formation of a Lysogen. (1) Bacteriophage lambda injects DNA through the bacterial cell wall. The resulting linear DNA molecule has sticky ends. **(2)** The DNA circularizes and **(3)** becomes ligated. At this point the phage has two choices. It can replicate its DNA, produce progeny phage, and kill the bacterium (*not shown*). Alternatively, it can integrate its DNA into the bacterial DNA **(4–6)** and remain quiescent for an indefinite number of bacterial generations. Both phage and bacterial proteins play important roles in the integration process.

PERSPECTIVE

Plasmids and phages are among the smallest and most efficient infectious agents in nature. Some are so efficient that they use the same nucleotide sequence to encode two different proteins. Plasmids have received considerable attention because of their medical importance: some plasmids carry genes that make their host bacteria resistant to antibiotics. Through our massive use of these drugs, we have encouraged the spread of plasmids to the point that almost every type of bacterium pathogenic to man now carries these infectious drug-resistance factors.

Phages are important in another way. For three decades molecular biologists concentrated on discovering how phages regulate their genes and replicate their DNA. As a result, phage studies provide the basis for most of our understanding of DNA. As our studies shift to complex organisms, plasmids and phages are assuming a new role in biology, that of a workhorse harnessed to purify genes.

More complex cells also contain small DNA molecules, some of which are considered to be plasmids and others to be viruses. Thus many of the principles discussed in this chapter apply to all cell types, and a variety of these small DNAs are now being used to manipulate genes in animal cells. Even the principle of lysogeny is important to us, for a version of that theme probably accounts for the ability of the AIDS virus to remain dormant for such a long time.

Questions for Discussion

1. What distinguishes a virus from a cell? Is a virus alive?
2. A bacteriophage can sometimes kill its host within 20 minutes, and in the course of infection 100 phage particles might be released from each cell killed. If you begin with a single phage particle in a culture of sensitive bacteria, how many phage particles would you have in 2 hours if you did not run out of bacteria?
3. Some bacterial viruses carry a gene for producing their own RNA polymerase. The promoters recognized by the viral polymerase differ from those used by the host RNA polymerase, thus providing the virus some control over its own destiny. If no RNA polymerase protein is carried in the virus particles, must the viral RNA polymerase gene have a promoter recognized by the host or viral RNA polymerase?
4. Devising ways to detect phages and plasmids is crucial to learn-

ing about them and to using them. Phages are generally detected by their ability to lyse bacteria, and their number is determined by counting plaques on a lawn of bacteria. Describe several ways to determine whether the cells in a bacterial colony contain a plasmid.

5. Figure 6-1 shows that single-stranded DNA can be distinguished from double-stranded DNA by electron microscopy. If two strands of a DNA molecule are not completely complementary when hybridized, they will not form a complete duplex, and the regions of noncomplementarity can be mapped by an electron-microscopic procedure called **heteroduplex mapping.** Figure 7-3 shows two plasmid molecules, one called cloning vehicle and the other called the recombinant DNA molecule. Suppose you cut both plasmids only once and at the same nucleotide sequence. This will produce linear DNA molecules. You next mix the DNA molecules, heat them to separate the strands, and then incubate them at the appropriate temperature to allow hybrids to form. You then examine the DNA in the electron microscope. Draw pictures of the types of structures you expect to see.

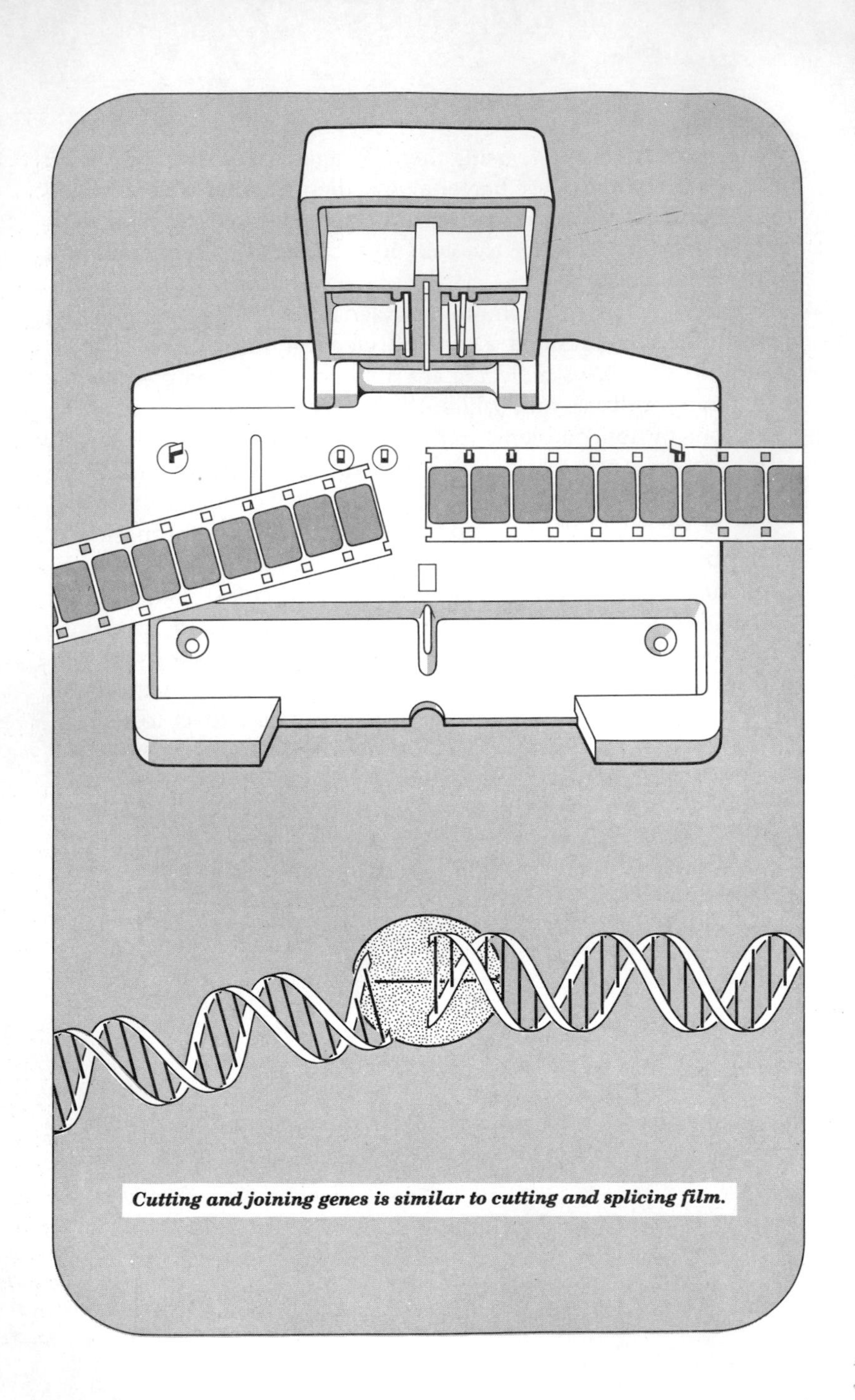

Cutting and joining genes is similar to cutting and splicing film.

CHAPTER SEVEN

GENE MANIPULATION

Cutting and Rejoining DNA Molecules in the Laboratory

Overview

Gene cloners move specific bits of genetic information from one DNA molecule to another by cutting and joining procedures that utilize specific enzymes. DNA is very long and contains a large number of sites at which cutting can occur. Consequently, cutting and joining often results in the creation of many different combinations of fragments. Biochemical methods based on the principle of complementary base pairing can be used to detect and locate a specific combination of rearranged fragments.

The enzymes that cut DNA are called *restriction endonucleases,* and they recognize and cut specific DNA sequences. Consequently, cutting DNA with these enzymes produces DNA fragments having discrete lengths. Enzymes can be obtained that recognize different sites in a DNA molecule, making it possible to cut DNA at a wide variety of locations. Cuts by different enzyme types produce DNA fragments having different sizes; two enzymes cutting the same DNA at the same time will cut in different places and will produce smaller fragments than will either enzyme cutting alone. It is possible to locate the cutting sites of one enzyme relative to another by comparing the sizes of the DNA fragments produced by the enzymes. This type of analysis produces a restriction map characteristic of the particular DNA being studied.

A mutation occurring in a DNA molecule at a site where a restriction endonuclease normally would cut will prevent the enzyme from cutting at that point; consequently, a larger DNA fragment will be produced when mutant DNA is cut. Other mutations can add restriction sites, making fragments smaller. It is now possible to detect some genetic diseases by analyzing the sizes of the DNA fragments produced by the cutting enzymes.

INTRODUCTION

Since the gross aspects of information organization in DNA are easily described by means of analogies between DNA and motion picture film, film metaphors are used below to begin describing gene manipulation. Imagine a film editor who feels that he can make an exciting, profitable, new motion picture by clipping out a scene from a John Wayne film and sticking it into a short Mickey Mouse cartoon. The spliced film will be longer than the original Mickey Mouse cartoon and, depending on which John Wayne scene was moved and where it was placed, the resulting movie will make more or less sense. A new motion picture will be created. In the same sense, gene "splicing" creates new genetic arrangements.

Like the film editor, a gene cloner needs two tools: scissors and splicing tape. The scissors biologists use are specific enzymes that obey an important rule: they cut only at specific places in the DNA. The specificity arises because the cutting enzymes recognize certain short sequences of nucleotides in the DNA. If this rule were applied to motion picture editing, cuts could be made only where specific events occur in the movie. For example, the editor might have to follow a rule according to which the film could be cut ONLY where someone laughs, but EVERY time anyone laughs in ANY movie, the editor MUST cut. Notice that the specificity rule leaves the editor with little control over the scissors; they cut automatically wherever they see a given sequence of events. If nobody ever laughs in a particular movie, the editor would not be allowed to cut that film; consequently, he would be unable to carry out splicing steps. Obviously, however, people may laugh frequently in a given movie. By the rule, such a film must be cut into many pieces. Consequently, an editor would have to sort through the many John Wayne film fragments to find the desired scene before he could begin splicing.

As an alternative strategy, the editor could splice each John Wayne fragment into a separate copy of the Mickey Mouse cartoon, creating a large number of short, new movies; he would then have to view each new movie to find the desired one. With film, the latter method is inefficient because an editor would have to make many nonproductive splices. He would then have to spend time screening the many new films to find the desirable one. Gene cloners, however, use this second strategy because they can easily create and examine thousands of "splicing" events and find a particular arrangement of nucleotides.

RESTRICTION ENDONUCLEASES

Restriction endonucleases are a group of enzymes that correspond to the scissors in the analogies developed above. There are many types, and the different types recognize different nucleotide sequences in DNA, often four or six base pairs long. The enzymes then cut both strands of DNA. Some enzyme types cut within the **recognition site** while other types cut close to it. In all cases the cut is at a specific site. In some cases the two DNA strands are not cut opposite each other—rather, the cuts are staggered. In the case illustrated in Figure 7-1, the cuts are offset by four nucleotides. Once the cuts in this example have been made, only four base pairs remain between cut sites. Under most conditions, four base pairs are not enough to hold DNA together, so the DNA molecule separates into fragments (Figure 7-1).

Many restriction endonucleases are commercially available. This collection of enzymes gives molecular biologists many different cutting options.

LIGATION

As pointed out above, some restriction endonucleases generate staggered cuts (Figure 7-1*b*). The four nucleotides in the single-stranded ends of the DNA molecules are complementary to the ends of other molecules generated by cutting with the same restriction endonuclease. Thus, when two DNA molecules having complementary ends collide, the single-stranded ends temporarily form base pairs, and under suitable conditions the two molecules will stick together (Figure 7-2). Hence the production of staggered cuts generates what molecular biologists call **sticky ends**. Chapter 5 mentioned DNA ligase, an enzyme that performs the essential function of joining DNA molecules together after DNA replication (step 2 in Figure 5-4). If this enzyme is present when two DNA molecules having sticky ends happen to come together, it will repair the breaks that had been introduced by the restriction endonuclease. Thus, DNA joining can be accomplished

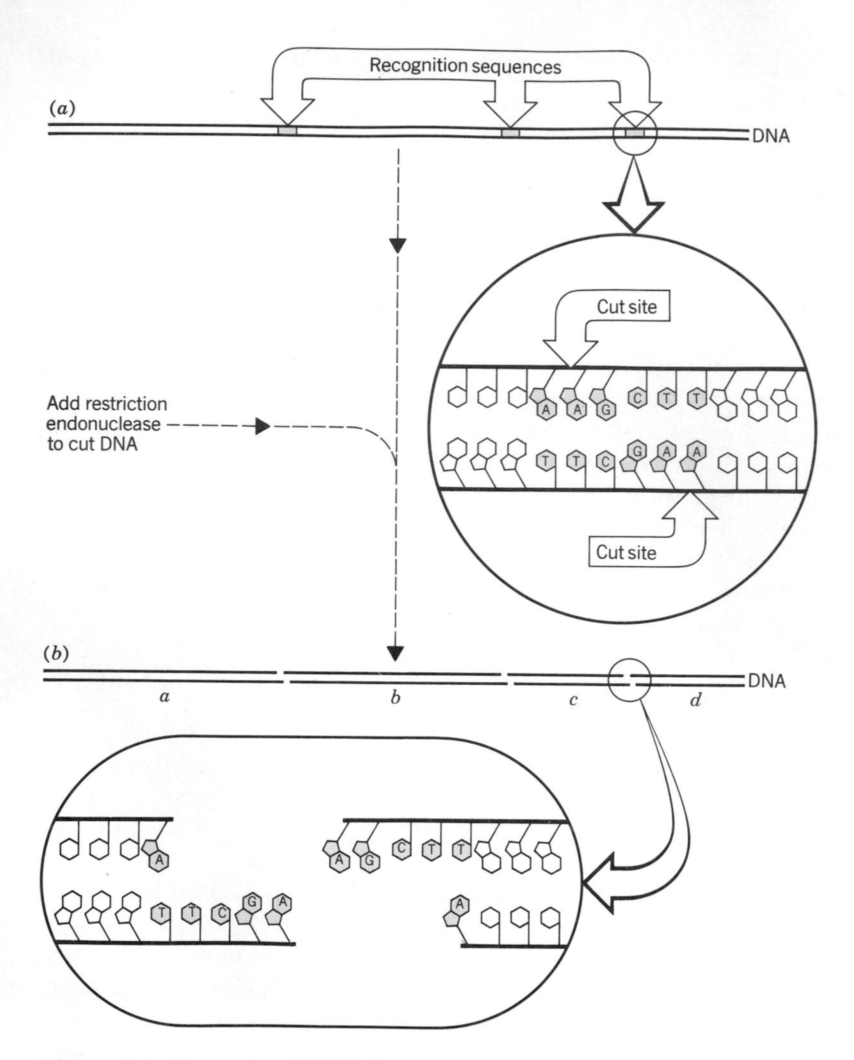

Figure 7-1 Cleavage of DNA by a Restriction Endonuclease. (**a**) A DNA molecule, depicted as two parallel lines, may contain many short nucleotide sequences recognized by restriction endonucleases. (**b**) When a restriction endonuclease is added to the DNA, it binds to the DNA and cuts it. Some of these enzymes produce staggered cuts. The DNA molecule in the example is converted into four shorter molecules, ***a***, ***b***, ***c***, and ***d***, each with "sticky ends" that can form base pairs with each other. For clarity, the DNA molecules are not shown as interwound helices, nor are hydrogen bonds between complementary base pairs shown.

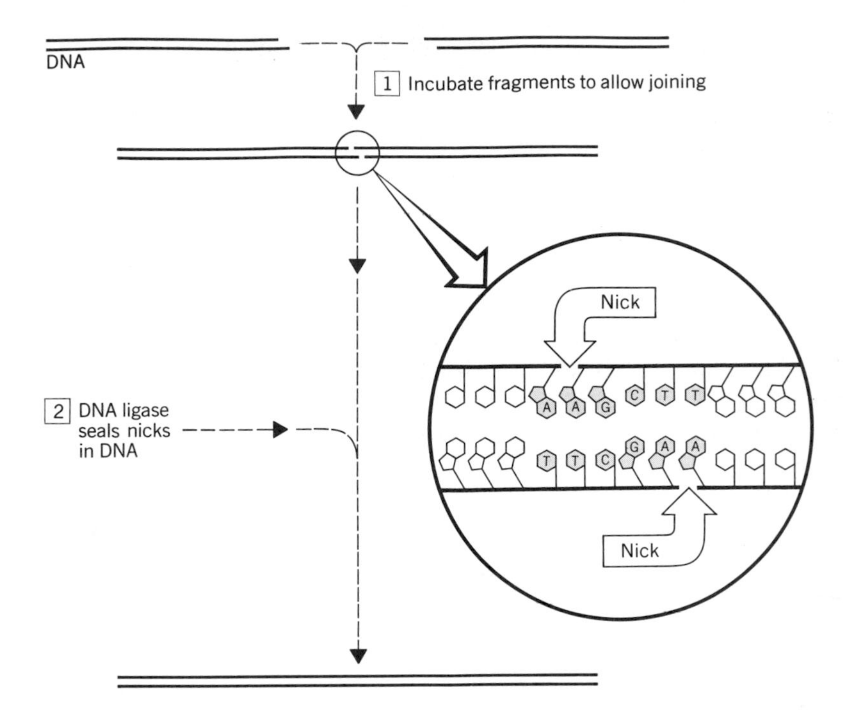

Figure 7-2 Joining Two DNA Fragments. **(1)** Two DNA molecules with complementary sticky ends are mixed and incubated. The molecules collide, and base pairs form. **(2)** The strand interruptions (nicks) are enzymatically sealed by DNA ligase.

by simply mixing together DNA molecules having complementary sticky ends and adding DNA ligase plus adenosine triphosphate **(ATP)** as a source of energy. Under the proper conditions, blunt ends can be ligated, a feature useful for joining ends created by restriction endonucleases that do not produce sticky ends.

CUTTING AND JOINING

The motion picture metaphor is useful when considering the details of cutting and joining DNA. First, imagine a large vat (representing a test tube) into which you have placed many, many unrolled copies of a particular John Wayne movie and a particular Mickey Mouse cartoon. The John Wayne film might represent many copies of an animal DNA, and the Mickey Mouse cartoon would represent many copies of a small DNA, the cloning vehicle, which can infect bacteria. Both are in solution in the same test tube. Now imagine that you and some friends enter the vat with splicing scissors and cut EACH film wherever anyone is shown laughing. This step corresponds to the addition of many identical restriction endonuclease molecules to the test tube to cut the DNAs. Soon, many fragments fill the vat. You and your friends leave the vat, so no more cutting can occur. The mixture is then stirred and the fragments collide. In the case of DNA in a test tube, some of the DNA collisions result in the ends sticking together, at least transiently. If DNA ligase is present, the fragments will become permanently joined.

As mentioned at the beginning of the chapter, the goal of gene "splicing" is analogous to inserting one specific scene from the chosen John Wayne movie into a COMPLETE Mickey Mouse cartoon. This chore can be simplified in two ways. First, before any cutting is done, tape the ends of the cartoon together to form a circle. Second, choose a cartoon in which someone laughs only once; then only one cut will occur in the cartoon. By starting with a circular cartoon, which can be cut only once, you ensure that the cartoon will always remain intact: no fragments will be created that must reattach to produce a complete cartoon. Eventually a John Wayne fragment will collide with one end of the cartoon, and the two can be joined. The remaining end of the cartoon will at some time collide with the free end of the John Wayne fragment already connected to the other end of the cartoon. When this second joining event occurs, a circle will have been formed, containing a complete cartoon plus a John Wayne fragment. If the two taped-together ends are released, the film can be projected from beginning to end. The new film will make sense except for a brief interruption where the John Wayne sequence has been inserted. This probably would not disrupt the cartoon any more than would a television commercial. A general scheme for DNA is outlined in Figure 7-3; in the case shown, the cloning vehicle occurs naturally as a circular plasmid.

In the vat, collisions between ends occur randomly, and much of the time the two ends of the cartoon simply rejoin. Likewise, John Wayne fragments attach to each other. In general, quite a mess is created. Occasionally, however, a single John Wayne fragment will collide with the cartoon. But since there are thousands of different John Wayne fragments, only rarely will any PARTICULAR fragment attach to the

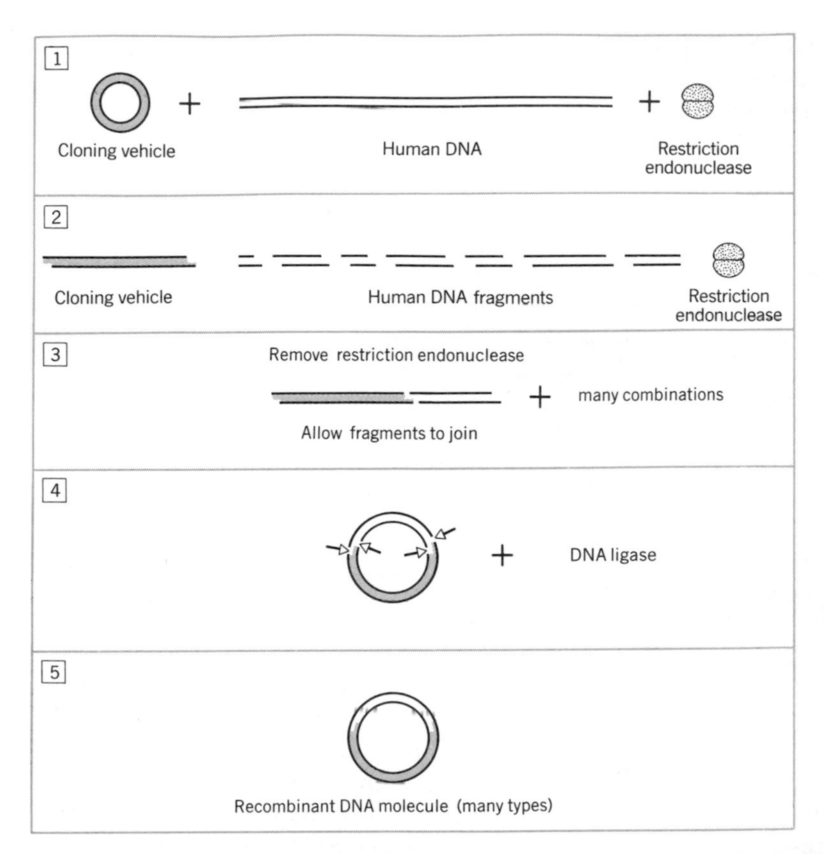

Figure 7-3 General Scheme for Forming Recombinant DNA Molecules. (**1**) Circular plasmid DNA (cloning vehicle), human DNA, and a restriction endonuclease are mixed. (**2**) Both DNAs are cut, producing a linear plasmid and many human DNA fragments. In this example, all DNAs in the mixture have complementary, sticky ends. (**3**) Occasionally a human DNA fragment will attach to one end of the plasmid. Many combinations form because many different types of human DNA fragment can join to the plasmid. (**4**) Eventually both ends of the human DNA fragment will have attached to the respective, corresponding ends of the plasmid. When DNA ligase is added, the discontinuities in the DNA strands (arrows in 4) will be sealed, producing a circular recombinant DNA molecule with no breaks in the DNA strands (**5**). The ligation of different human DNA fragments to plasmids causes the formation of many types of recombinant molecules.

cartoon. Thus, in genetic engineering, trillions of DNA molecules must be incubated together before there is a reasonable chance that the desired fragment will attach to the cloning vehicle.

RESTRICTION MAPS

Restriction endonucleases occur naturally in a large number of different bacterial species. These enzymes appear to be part of the natural defense mechanism protecting bacterial cells against invasion by foreign DNA molecules such as those contained in viruses. A crucial element to this protective device is that the **nuclease** must discriminate between its own DNA and the invading DNA; otherwise the cell would destroy its own DNA. The recognition process involves two elements. First there are specific nucleotide sequences that act as targets for the nuclease. Second, the cell is able to place a protective chemical signal on all the target sequences that happen to occur in its own DNA. The signal modifies the DNA and prevents the nuclease from cutting. Invading DNAs would lack the protective signal and would be chopped up by the nuclease. Restriction endonucleases from different bacteria often recognize different target sequences. Thus restriction endonucleases purified from different bacteria can be used as enzymatic tools to cut DNA at different, specific sites.

In addition to their role in DNA cloning procedures, restriction endonucleases play an important part in the analysis of nucleotide sequences of DNA molecules (described in Chapter 9). The initial step in these analyses is to construct a **restriction map**; a procedure is outlined below to provide familiarity with these cutting enzymes. First, return to the analogy between DNA and motion picture film, and imagine that you wish to study a film loop 1000 feet long. You are not allowed to use a movie projector, and the film has no ends: it is a circle. You could begin to examine this tangled mass if you could cut the film at SPECIFIC places to produce DISCRETE, manageable pieces. Now suppose that you have three kinds of scissors: type A, which cuts when somebody laughs; type B, which cuts when a dog bites a man; and type C, which cuts whenever a car door is opened (these scissors correspond to different restriction endonucleases cutting at specific sites on DNA). Assume in the film you are studying that somebody laughs only once. Therefore type A scissors will cut only once, producing two ends and giving you a reference point:

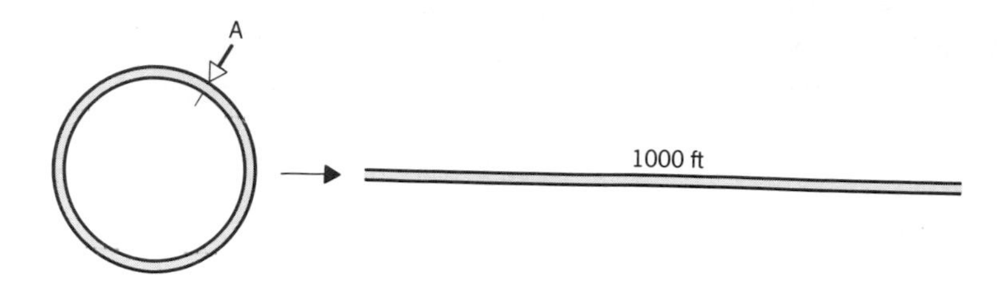

If in your film men are bitten by dogs three times, the circle will be cut into three fragments by type B scissors. You can measure the lengths. Suppose that the fragments are 100, 300, and 600 feet long.

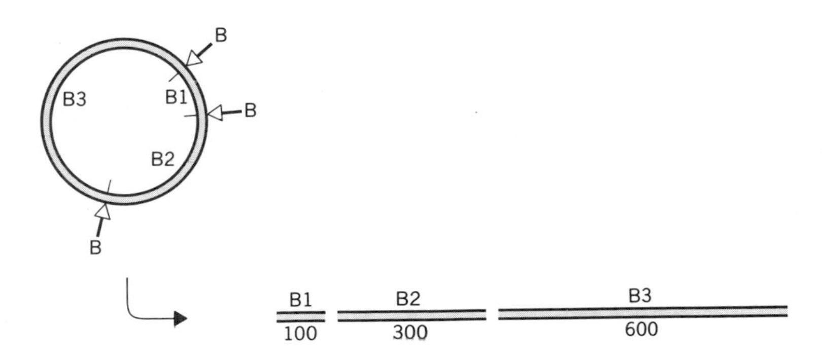

Likewise, if a car door opens twice in the film, type C scissors will cut two times, producing two fragments. Suppose that these fragments are 200 and 800 feet long.

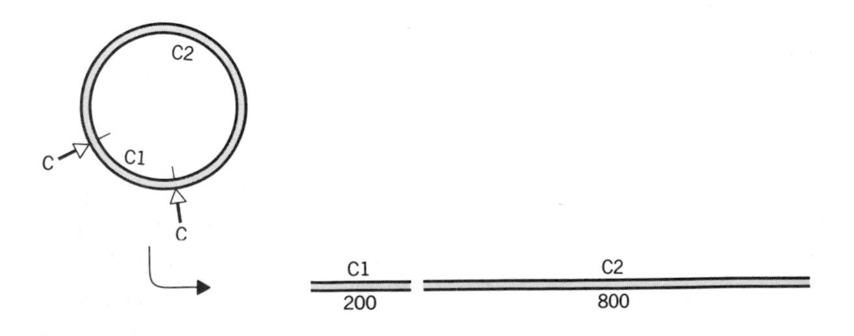

Even smaller fragments can be obtained by cutting with more than one scissors type. For example, type A and type B combined should produce four fragments. Suppose that you find the resulting lengths to be 50, 100, 300, and 550 feet. In such a case the type A cut must have

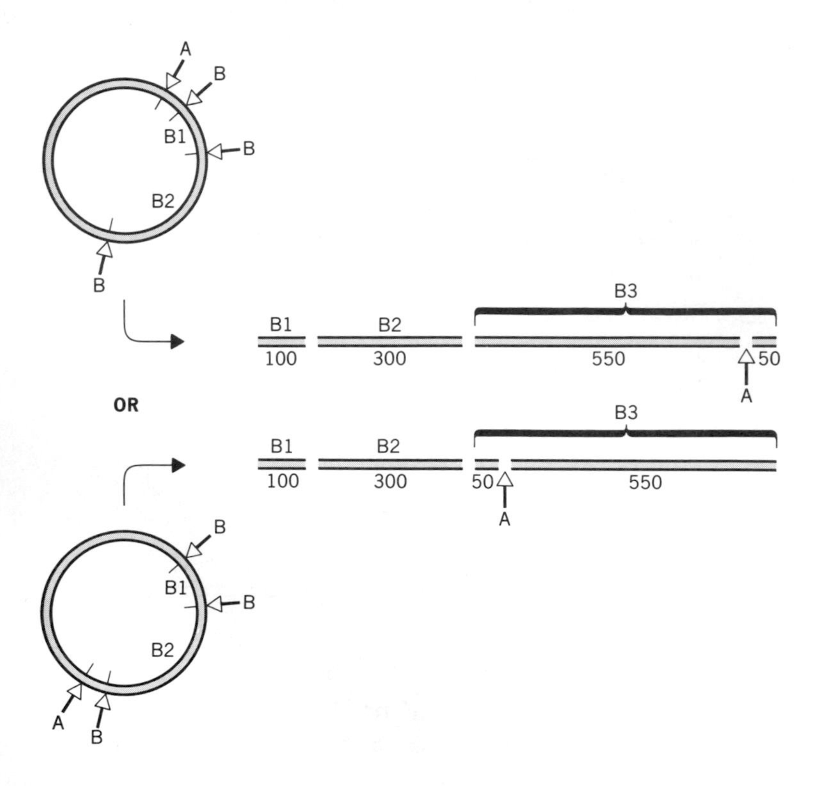

occurred within the 600-foot fragment created by the type B enzyme (B3), producing two new fragments (50 and 550). There are only two ways the film can be arranged to produce this result, and in both cases the laugh (A) occurs between the two most widely spaced dog bites (B). Thus the map is beginning to develop. At this point you cannot tell whether cut A is nearer to fragment B1 or to fragment B2.

The next combination involves cutting with type A and type C scissors. This time you find three fragments, which are 200, 375, and 425 feet long. The type A cut must have occurred within the 800-foot C2 fragment, generating two new pieces 375 and 425 feet long. Again there are two places where the A cut could be relative to the two C cuts.

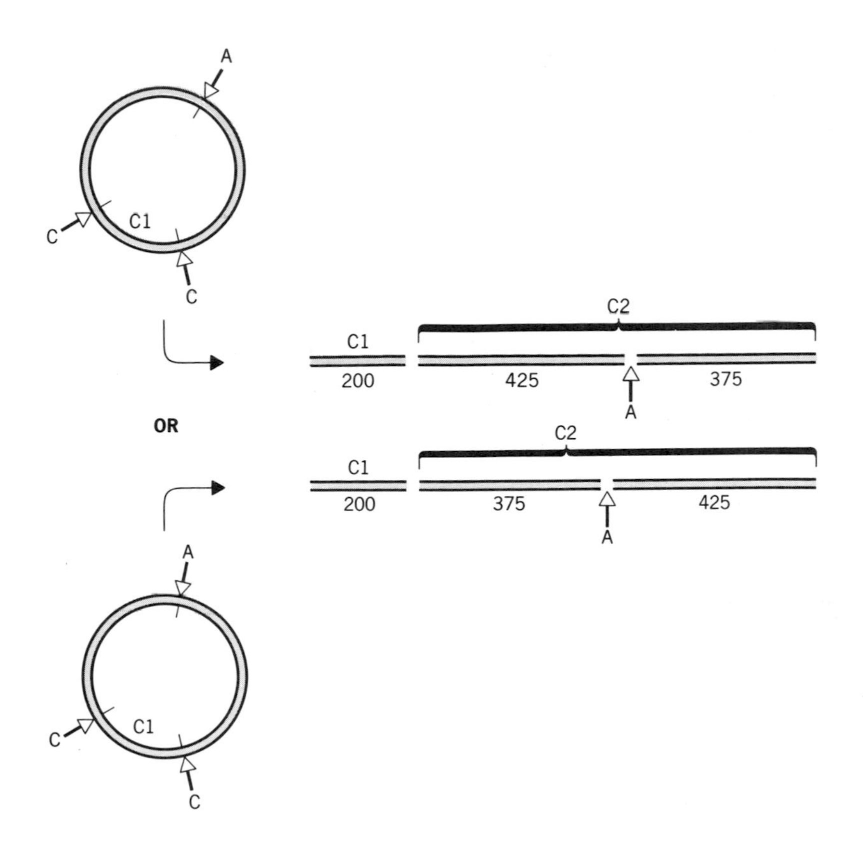

You can also cut with a combination of type B and type C enzymes. Suppose this produces five pieces, 75, 100, 125, 225, and 475 feet long. The 100-foot piece is B1, but B2 (300 feet) and B3 (600 feet) have disappeared. This must mean that one C-type cut occurs in B2 and the other in B3. The 75- and 225-foot fragments add up to 300 feet, which corresponds to B2, so one C-type cut is 75 feet from one end of B2. The sum of 125 and 475 is 600, the value of B3; thus the other C-type cut is 125 feet from one end of B3. Since the two C-type cuts are either 200 or 800 feet apart on the accompanying circular map, there is only one way the map can fit together.

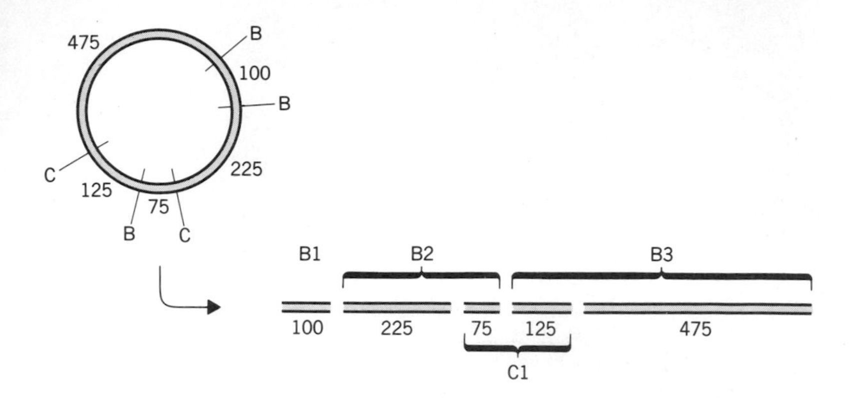

By adding the results from the A + C, the A + B, and the B + C combinations, it is possible to determine the position of the A cut. Since A is in B3, only 50 feet from a B cut, and far from the C cuts, A must map as follows:

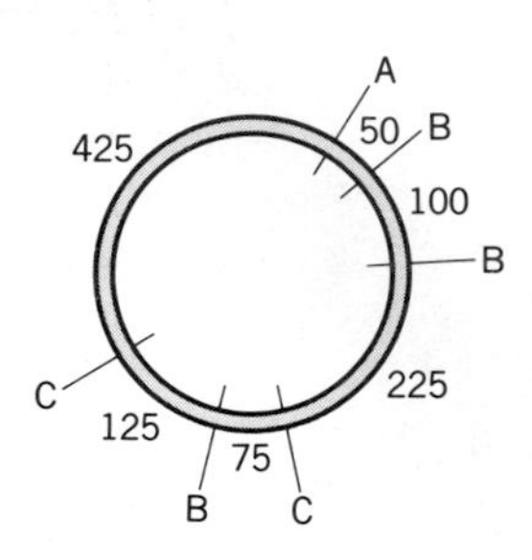

When all three types of cut are made simultaneously, six fragments will be produced.

B1 B2 B3
100 225 75 125 425 50
C1 A

At this stage in the film analogy one would say that the sequence of events, beginning at A, is laugh, two dog bites, a car door opening, another dog bite, and finally another car door opening. The film distance between the events is also known.

In the case of DNA, the size of each of the fragments is measured by a technique called **gel electrophoresis**. A semisolid material such as agar is first poured and allowed to solidify to form a slab. A small well is formed in the gel, and a mixture of DNA fragments, produced by restriction endonuclease-induced cleavage, is placed in the well. An electric field is then applied across the gel, forcing the negatively charged DNA molecules to move through the gel. DNA molecules having the same size move together as a group. If trillions of identical DNA fragments are present, you can see them as a band by staining the gel. Smaller DNAs move faster through the gel than larger ones; consequently, fragments of different lengths produce bands at different positions in the gel (Figure 7-4). The size of DNA molecules in each band is determined by comparing how far a given band moved into the gel relative to bands of DNA molecules whose lengths are already known.

DNA in a band can be removed from the gel, inserted into a cloning vehicle, transferred into a bacterium, and amplified by growth of the bacterium. The nucleotide sequence can then be determined as described in Chapter 9. By knowing the map order of the fragments (using the type of logic just outlined), it is possible to fit together the short nucleotide sequences from all of the fragments like a jigsaw puzzle, eventually obtaining the nucleotide sequence of the entire DNA molecule.

Restriction mapping is also used as a diagnostic tool for prenatal detection of certain genetic diseases. In sickle-cell anemia, a single nucleotide change is responsible for the disease (see Figure 5-5). The base change occurs in a restriction endonuclease recognition site, and the site is so altered by the change that it is no longer recognized by the enzyme. Thus, the enzyme does not cleave at that location. Consequently, the restriction map of DNA from a sickle-cell fetus differs from that obtained from normal DNA. The principle is illustrated in Figure 7-5 using the restriction mapping example given earlier in this chapter. In this example, the base change causing the disease occurs in the recognition site for the type A restriction endonuclease, eliminating that site from the DNA. The map of the DNA from the diseased fetus would then be easily distinguishable from that of a normal fetus: two normal fragments (425 and 50) would be replaced by a new frag-

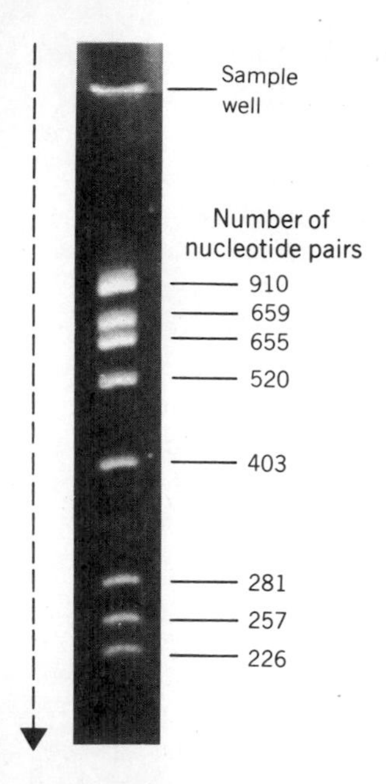

Figure 7-4 Gel Electrophoresis Display of Restriction Fragments of DNA. A mixture of DNA fragments was placed in a sample well (slot) in a gel and driven into the gel by an electric field. Smaller fragments moved faster than larger ones, so when the electric field was turned off, the fragments had separated. The gel was stained to reveal the DNA bands and then photographed. Each band represents many DNA molecules having the same length. The direction of DNA movement is from top to bottom; the number of nucleotide pairs in each fragment is indicated at right for each fragment. (Photograph courtesy of Richard Archer, Lasse Lindahl, and Janice Zengel, University of Rochester.)

ment (475) in the mutant DNA. Restriction fragment changes are usually detected by a method called Southern hybridization (Chapter 9).

There are many regions in human DNA that vary from one person to another such that the number of nucleotides, the length of the DNA, between two particular spots cut by a restriction endonuclease differs from person to person. This is sometimes due to short sequences being repeated different numbers of times, sometimes to sequences being added or substrated from the region, and sometimes to a restriction site being created or eliminated in the region. Such regions that have different forms are called **restriction fragment length polymorphisms** or **RFLPs** (polymorphism means multiple forms). These

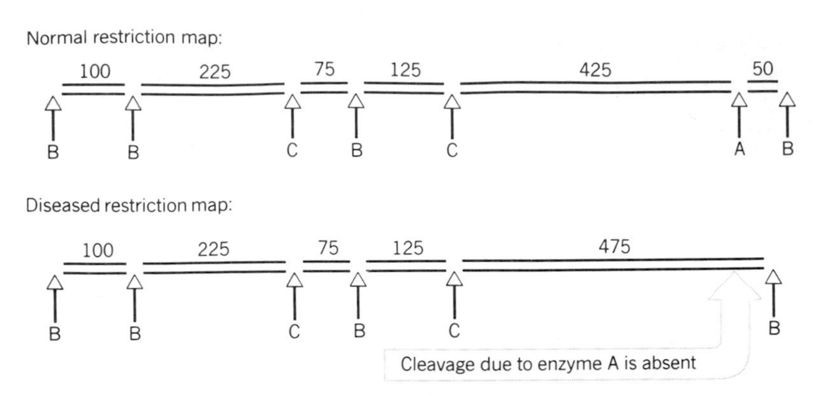

Figure 7-5 Diagnosing a Genetic Disease by Restriction Mapping. In this hypothetical example, a change in the nucleotide sequence in DNA causes a disease and also eliminates a known restriction site (site for enzyme A). Thus DNA from a diseased cell has an altered restriction map. Cells from higher organisms such as humans have two copies of each DNA molecule, one derived from the mother and one from the father. Although these molecules are never identical over their whole length, they may be identical in the region being examined. If in this region DNA molecules derived from the parents are both normal or both pathogenic, the analysis produces the results shown in these two diagrams. If, however, the DNA from one parent is normal but the DNA from the other parent is pathogenic, a mixture of the restriction fragments is observed. Sickle-cell anemia is known to be due to a single nucleotide change in DNA (see Figure 5-5), and it is now possible to diagnose the disease by examining restriction maps of fetal DNA.

regions do not necessarily contain genes, and the function of most of them in the human **genome** is unknown. By carefully examining the DNA of members of families that carry genetic diseases it has been possible to find forms of particular RFLPs that tend to be inherited with a particular disease. Generally the fragment differences are not due to a restriction site being created or disrupted by the diseased state itself (sickle-cell anemia is an exception), but rather the nucleotide sequence differences just happen to be near the gene involved. By being close to the gene, a particular form of a polymorphism tends to stay with the diseased gene during the chromosomal rearranging and sorting that occurs prior to fertilization and formation of the fetus. Thus RFLPs serve as markers of disease, and they are now useful for detecting diseases such as Huntington's disease, cystic fibrosis, sickle-cell anemia, hemophilia, and certain types of colon and lung cancer.

Restriction analyses are also used to track down the source of infectious diseases. Hospital infections caused by the pathogenic bacterium *Staphylococcus aureus* represent a good example. Outbreaks of "hospital *Staph*" often occur in hospital nurseries and cardiac units, and because these strains are often resistant to many antibiotics, they have become a very serious problem. Locating and eliminating reservoirs of the organism is a major challenge. In one case, restriction fragment analyses of bacteria taken from surgical patients suffering from post-operative toxic shock syndrome, one of the consequences of infection by *S. aureus*, showed beyond doubt that the surgeon had been harboring the bacteria in his nose. Once this became known, steps were taken to rid the surgeon of his symptom-less infection.

VISUALIZATION OF CLONED DNA

Most of the basic tools used in gene cloning have now been introduced: the restriction endonucleases that cut DNA molecules in specific places, the ligases that join DNA molecules, the plasmids and phages that carry new combinations of DNA molecules into cells where they can be reproduced, and the nucleic acid probes used to identify specific colonies or plaques that contain cloned genes. The next step is to obtain physical evidence showing that new DNA has actually been inserted into a phage or plasmid.

One type of evidence for insertion involves changes in restriction maps. Figure 7-6 illustrates how a human DNA fragment changes the

Figure 7-6 Analysis of Recombinant DNA by Gel Electrophoresis. (1) Plasmid and human DNA molecules are cut and joined together to form recombinant plasmid DNA molecules. Many combinations of DNA fragments join, producing many different recombinant DNAs. **(2)** The recombinant DNAs are introduced into bacterial cells, and individual colonies are grown. A bacterial colony containing cloned DNA is identified (Chapter 8), and recombinant plasmid DNA is isolated from it. (All of the recombinant DNA molecules from a colony are identical. They all have two restriction endonuclease cleavage sites, one at each junction between human and plasmid DNA.) **(3)** Cleavage of the recombinant plasmid DNA with restriction endonuclease produces two discrete DNA fragments. **(4)** The original plasmid contains only one restriction endonuclease cleavage site, so it is cleaved only once by the nuclease. **(5)** The products of steps 3 and 4 are analyzed by gel electrophoresis, as described in Figure 7-4. **(5)** The original plasmid produces only one band; the recombinant plasmid produces two.

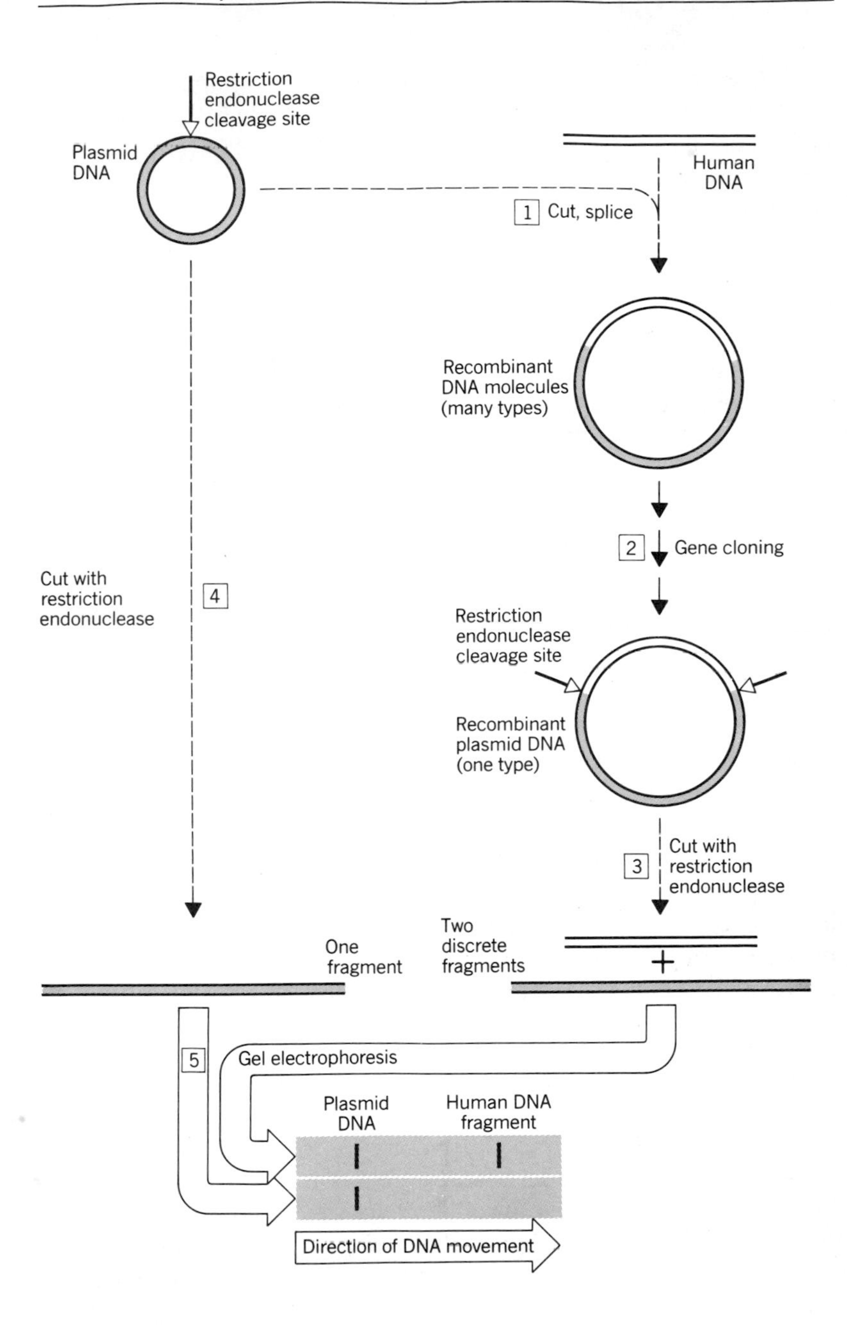
Restriction
endonuclease
cleavage site
Plasmid
DNA
Human
DNA
1 Cut, splice
Recombinant
DNA molecules
(many types)
2 Gene cloning
Cut with
restriction
endonuclease
4
Restriction
endonuclease
cleavage site
Recombinant
plasmid DNA
(one type)
3 Cut with
restriction
endonuclease
One
fragment
Two
discrete
fragments
+
5
Gel electrophoresis
Plasmid
DNA
Human DNA
fragment
Direction of DNA movement

restriction map of the plasmid into which it has been cloned. In this example, the plasmid contains a single site where a particular restriction endonuclease cuts. At this site human DNA fragments, generated by the same type of endonuclease, are "spliced" into the plasmid molecules to form recombinant DNA molecules (step 1, Figure 7-6). The recombinant DNA molecules are introduced into bacterial cells, individual clones are obtained (see Chapter 8), and recombinant DNA molecules are purified. This produces a single type of recombinant plasmid (step 2). When these DNAs are cut with the restriction endonuclease (step 3), two pieces of DNA are created. These two pieces can be analyzed and compared with the single piece generated by cleaving the original plasmid (step 4), using gel electrophoresis (step 5).

PERSPECTIVE

DNA molecules are so long and contain so much information that until the discovery of restriction endonucleases there seemed to be little hope of determining extensive nucleotide sequences. Now that DNA molecules can be cut into discrete, manageable fragments, determining nucleotide sequences has become routine. Attention is currently focused on determining the complete nucleotide sequence of the human **genome**, an effort comparable in scope, scale, and funding to that required to put a man on the moon. The next chore will be to discover what the various sequences do. Some information can be obtained by examining how specific, cloned sequences function when placed inside living cells, especially if the sequences encode proteins. But the task becomes massive when one begins to ask detailed questions about how different regions of DNA interact to coordinate the control of gene expression.

Restriction fragment analyses have many practical applications. Already in place are tests to identify fetuses with genetic diseases and adults who may be predisposed to these conditions. Such analyses can also be used in general identification processes, and law enforcement agencies are taking advantage of the technology. More controversial issues, just now coming into the political arena, include screening for insurance coverage and employment suitability.

Questions for Discussion

1. The molecular scissors used to cut DNA are called restriction endonucleases. What does endonuclease mean? (Define the word fragments *endo-*, *nucle-*, and *-ase.*)
2. Restriction endonucleases are extremely specific in terms of their nucleotide recognition sequence. How does their natural function in cells help explain why they are so specific?
3. Suppose that you are studying a new plasmid that is 2500 base pairs long, and you wish to construct a restriction map. You treat the plasmid DNA with a set of restriction endonucleases and measure the size of the resulting DNA fragments by gel electrophoresis. You obtain the following results (the abbreviations refer to specific restriction endonucleases; the numbers are the sizes of the DNA fragments in base pairs):

   EcoRI — 2500
   HindIII — 2500
   PstI — 2500
   MboI — 1300, 800, 400
   MboI + *EcoRI* — 1300, 600, 400, 200
   MboI + *HindIII* — 1300, 800, 300, 100
   MboI + *PstI* — 1000, 800, 400, 300
   EcoRI + *HindIII* — 2000, 500
   EcoRI + *PstI* — 1600, 900
   HindIII + *PstI* — 2100, 400

   Construct a map based on the information given above. In your map place base pair 1 at the *HindIII* site.

4. Suppose that you had two restriction endonucleases, *EcoRI* and *MboI*, each in a separate test tube. Somehow the labels came off the tubes. How could you determine which tube contained *EcoRI* using the plasmid described in question 3?

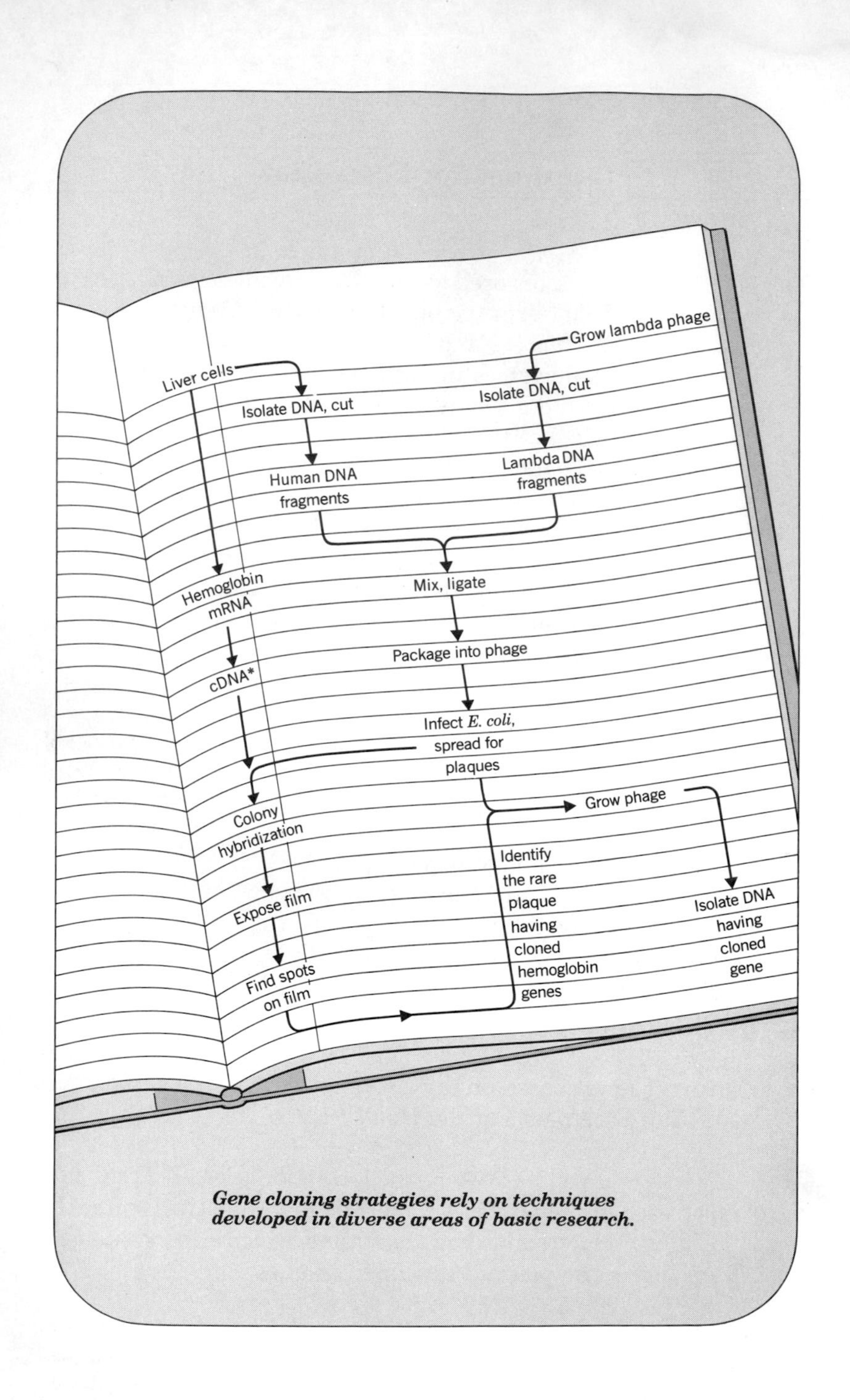

Gene cloning strategies rely on techniques developed in diverse areas of basic research.

CHAPTER EIGHT

CLONING A GENE

Isolation of a Hemoglobin Gene

Overview

There are now many strategies for cloning genes. In each case small fragments of DNA are inserted into cloning vehicles (i.e., into plasmid or viral DNA molecules). The vehicles carry the fragments into living cells, where the vehicles and their attached fragments reproduce. Cells containing plasmids grow into individual colonies while those infected with a virus generate individual plaques. Both colonies and plaques can be tested for the presence of a particular piece of DNA by nucleic acid hybridization using a radioactive probe.

Gene cloning strategies with plasmids often use antibiotic-resistance properties to help locate bacterial colonies into which genes have been cloned. One type of strategy employs plasmids having genes for resistance to two different antibiotics. One antibiotic-resistance gene contains a restriction endonuclease cleavage site at which DNA fragments can be inserted; insertion of a fragment into this site on the plasmid destroys that resistance gene. The second antibiotic-resistance gene is not affected. Thus recombinant plasmid DNA molecules will confer resistance to the second antibiotic only. When bacterial cells are transformed with plasmids thought to be recombinants, colonies are obtained that grow on agar containing the second antibiotic but not on agar containing the first antibiotic. Such colonies are then tested for the presence of a specific gene, often by nucleic acid hybridization.

INTRODUCTION

As pointed out earlier, gene cloning is much like baking a cake; recipes are available for both processes. This chapter describes some general cloning strategies to tie together the concepts developed in the preceding chapters. The focus is on the hemoglobin genes, the first mammalian genes to be cloned. Hemoglobin is the blood protein responsible for moving oxygen from our lungs to our body tissues, and a number of genetic disorders have been associated with defective hemoglobin. Among these is the serious disease called sickle-cell anemia. The hemoglobin genes have also elicited interest because they code for separate proteins that function at different stages of our lives. Biologists want to understand how one hemoglobin gene is switched on and how another is switched off. Having the hemoglobin genes cloned makes many new experiments possible. An outline of the procedures is sketched in Figure 8-1.

OBTAINING DATA

The first step in gene cloning is to obtain DNA that contains the gene of interest. The initial hemoglobin studies were done with rabbits, but similar procedures would work for humans. With few exceptions, all the cells in a rabbit contain identical DNA molecules. Thus DNA from most rabbit cells can serve as a source for hemoglobin genes, and almost any type of body tissue can be ground up to obtain these genes. The original recipe called for extracting DNA from rabbit livers. Liver tissue was frozen, placed in a blender, and chopped until the cells were

---

Figure 8-1 Outline of Cloning Procedure for Obtaining Genomic Clones of Hemoglobin DNA. Purified hemoglobin mRNA (left portion of diagram) is converted to a DNA form by using the enzyme reverse transcriptase. The DNA copy, called complementary DNA (cDNA), is inserted into a plasmid vector, and bacteria are transformed with recombinant plasmids containing cDNA. Bacteria containing the appropriate recombinants are selected, and large amounts of cDNA plasmid are isolated. This plasmid DNA is cut and radioactively labeled to use as a probe.

Rabbit DNA (right portion of diagram) is fragmented, and the fragments are joined to bacteriophage DNA. Phage proteins are added to package the recombinant DNA into phage particles. Bacteria are then infected with the recombinant phage, and plaques form. Those plaques that contain hemoglobin DNA are identified by hybridization using the radioactive cDNA plasmid as a probe.

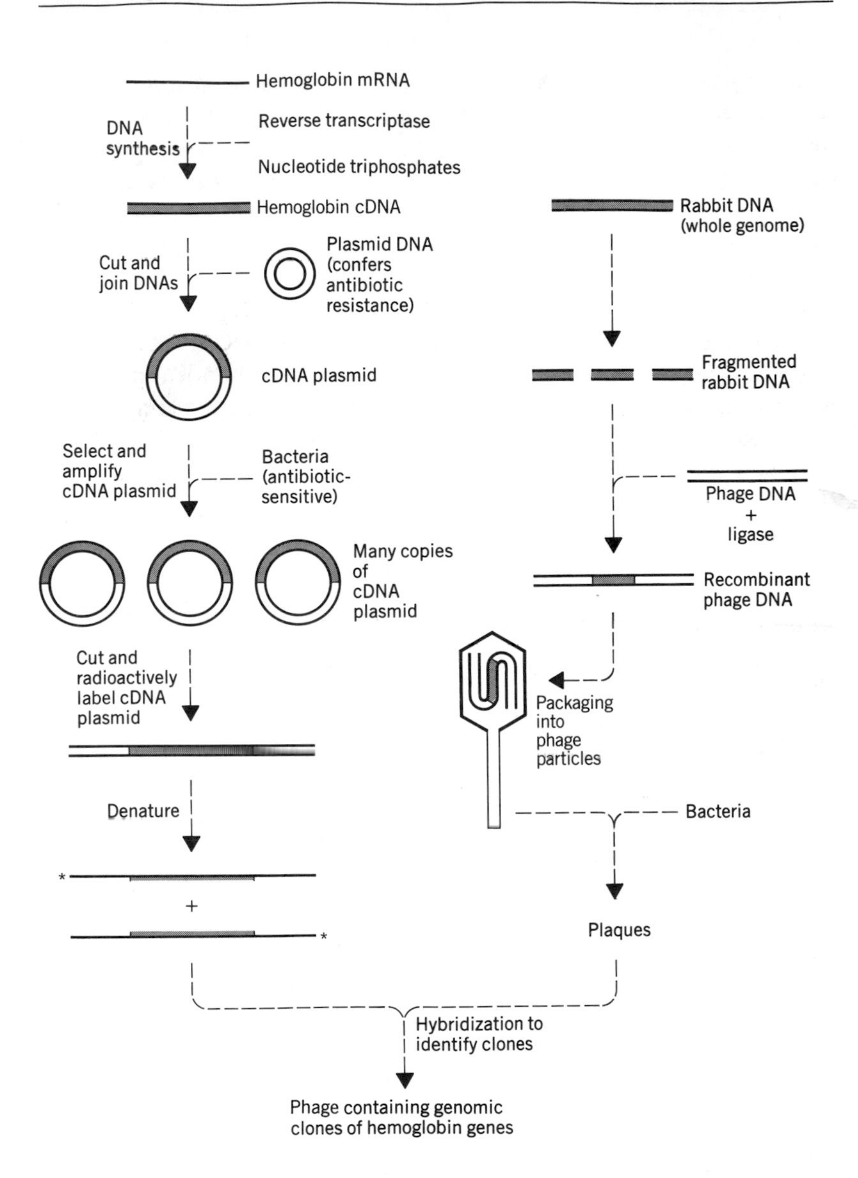
Hemoglobin mRNA
DNA synthesis
Reverse transcriptase
Nucleotide triphosphates
Hemoglobin cDNA
Cut and join DNAs
Plasmid DNA (confers antibiotic resistance)
cDNA plasmid
Select and amplify cDNA plasmid
Bacteria (antibiotic-sensitive)
Many copies of cDNA plasmid
Cut and radioactively label cDNA plasmid
Denature
*
+
*
Rabbit DNA (whole genome)
Fragmented rabbit DNA
Phage DNA + ligase
Recombinant phage DNA
Packaging into phage particles
Bacteria
Plaques
Hybridization to identify clones
Phage containing genomic clones of hemoglobin genes

broken, releasing the molecules held inside. A detergent solution was then added to unfold cellular proteins, and the cell lysate was treated for several hours with a **protease**, an enzyme that fragments proteins. Such treatment removes proteins that may be sticking to the DNA. It also inactivates nucleases present in the extract that might otherwise begin cutting the DNA. Next an oily substance called **phenol** was added to the lysate. Phenol and the watery cell lysate do not mix. However, if the combination is shaken vigorously, many of the proteins that might have escaped enzymatic digestion move into the phenol. When the mixture is allowed to stand, the phenol and the cell lysate form two layers in a test tube. The DNA-containing layer can be removed and saved. By this stage most of the cellular proteins have been removed from the DNA preparation. The rabbit DNA in the lysate was further purified by centrifugation in a density gradient as described in Chapter 6. The DNA formed only a single band in the centrifuge tube, and it was easily removed with a syringe and placed in another test tube. Hemoglobin genes, millions of them, were in the test tube. But so were millions of copies of a hundred thousand other genes.

CLONING INTO BACTERIOPHAGE LAMBDA DNA

Hemoglobin genes were initially separated from all the other genes in the DNA preparation by cloning them into bacteriophage lambda DNA (see Chapter 6 for description of bacteriophages). The phage is particularly attractive because large DNA segments can be inserted into its DNA; thus the resulting clones might contain several adjacent genes as well as nearby noncoding regions. In this part of the recipe lambda phage particles were first purified by a method similar to that described in Chapter 6 (Figure 6-6). Next, DNA was extracted from the phage particles, using many of the same steps described above for purification of rabbit DNA. To make room for the rabbit DNA inside the virus particles, the lambda DNA was shortened (not all of the information in lambda DNA is required for the phage to infect bacterial cells). The phage DNA was cut into several pieces by a restriction endonuclease, and fragments lacking information essential for infection were discarded. Rabbit DNA was broken into large pieces and mixed with the phage DNA. Ligase was added to join the DNA fragments. Most of the resulting DNA molecules contained only phage DNA, but often a piece of rabbit DNA had been added. The DNA molecules were then coated with phage proteins, neatly packaging the DNA inside phage particles. These phages formed *in vitro* (i.e., in a test tube) were next allowed to infect *E. coli* cells, which subsequently were spread on agar. This spreading step separated the phage particles from each other. Phage plaques formed in the lawn of bacteria as each phage reproduced (Figure 6-5). Every plaque arose from a different phage particle, and each plaque contained millions of identical phage particles. Unfortunately, very few contained hemoglobin genes.

At this point the problem was to locate the rare plaques that contained hemoglobin genes. For some genes it is possible to look for specific properties of the plaques arising from recombinant proteins produced by the infected bacterial cells. For example, some proteins catalyze chemical reactions that cause a plaque to become a particular color if the protein is present. Frequently this is not the case, however, and nucleic acid probes or specific antibodies must be used to identify plaques that contain the genes being sought. Antibodies are described more extensively in a later chapter and are not dealt with further here. In methods using nucleic acid probes, DNA in each plaque is tested for its ability to hybridize with a nucleic acid complementary to the gene or genes being sought, relying on the principle of complementary base pairing as discussed in earlier chapters (see Figure 5-10). Generally, nucleic acid probes are radioactively labeled to facilitate detection.

OBTAINING NUCLEIC ACID PROBES

An appropriate nucleic acid probe can be obtained easily if the protein product of the gene being sought has already been purified. The amino acid sequence for part of the protein is determined, and the genetic code (Figure 3-6) can be used to derive a short DNA sequence likely to be part of the gene. An oligonucleotide having that sequence is then synthesized and radioactively labeled.

Another source for a radioactive probe is messenger RNA from the gene being sought. In most cases of gene cloning it is very difficult to isolate mRNA that represents the information from just a single gene.

Generally, cells make many types of messenger at the same time, and the mRNA molecules are chemically and physically too similar to be separated. However, hemoglobin represents a special case, for the mRNA in red blood cells is mainly hemoglobin mRNA. Thus rabbit blood was collected, and the red blood cells were concentrated by centrifugation, forming a pellet in the bottom of a test tube. The pellet of cells was resuspended in a small volume of water and salts, and the cells were broken by adding detergents. The resulting lysate was extracted with phenol to help remove proteins, and alcohol was added. Alcohol caused the RNA to form a white precipitate, which was separated from other cell components by centrifugation. As a further precaution, the RNA was dissolved in water and centrifuged in a density gradient. RNA has a unique buoyant density and is readily purified by this method. At this stage the sample still included ribosomal and transfer RNA as well as hemoglobin mRNA.

Hemoglobin mRNA was separated from other types of RNA by taking advantage of a feature unique to many kinds of mRNA found in higher organisms: these RNAs often have several hundred adenines attached to one end. A glass column, similar to that illustrated in Figure 5-9, was filled with cellulose to which single-stranded DNA, composed only of thymidines, was attached firmly. The RNA mixture was passed through the column. As the RNA percolated through, the stretches of adenines on the hemoglobin messengers formed complementary base pairs with the long runs of thymidines fixed to the cellulose. The complementary base pairs were strong enough to prevent hemoglobin mRNA from flowing through the column. The other RNA molecules did flow through, and they were discarded. Hemoglobin mRNA was then removed from the column by breaking the base pairs, which can be achieved by warming the column gently.

At this stage the mRNA was almost ready to use to identify a phage plaque containing a hemoglobin gene. But first the information in it, the sequence of nucleotides, had to be converted to a highly radioactive form. This could have been done by mixing together RNA, radioactive nucleotides, and a type of DNA polymerase called **reverse transcriptase**. This polymerase is obtained from **retrovirus** particles, and it makes DNA from free nucleotides, using the RNA as a template. The DNA, called **complementary DNA (cDNA)**, can be made highly radioactive if during its synthesis radioactive nucleotides are present in reaction mixtures.

CLONING WITH PLASMIDS

In many examples of gene cloning, radioactive cDNA is suitable to locate plaques containing cloned genes. In the initial cloning studies with hemoglobin, however, it was necessary to screen hundreds of thousands of phage plaques, requiring huge amounts of cDNA. It was decided to first clone the cDNA into a plasmid. Then large amounts of it could be obtained easily from bacterial cells. This cloned cDNA would be radioactively labeled and used to test the phage plaques for hemoglobin genes. The plasmid methods initially used for cloning hemoglobin cDNA are now seldom utilized and will not be described. Instead, a more general strategy is presented.

DNA fragments from the organism being examined (in our case rabbit hemoglobin cDNA) are inserted into plasmid DNA using restriction endonucleases and ligase, the DNA mixture is added to *E. coli* cells, and some of the DNA molecules enter the bacteria. The task is to isolate a pure culture of bacterial cells harboring a plasmid that contains only a particular piece of rabbit cDNA (hemoglobin cDNA). After the bacteria have taken up DNA, they continue to grow and divide, and soon there may be billions of bacteria in the culture. The cultured *E. coli* cells can be divided into four classes:

1. *E. coli* that failed to take up any plasmid DNA.
2. *E. coli* that took up plasmid DNA without any rabbit cDNA inserted.
3. *E. coli* that took up plasmid DNA with rabbit DNA inserted but not the particular rabbit gene being sought.
4. *E. coli* that took up plasmid DNA into which the desired rabbit gene was inserted.

The fourth category is the important one, and members of this category are generally very rare.

Two tricks are used to increase the odds for finding plasmids with rabbit genes. First, the plasmid chosen as a cloning vehicle contains a gene for resistance to **tetracycline**, a drug that prevents bacterial growth. Thus, the antibiotic can simply be added to a flask seeded with the bacteria so that only cells protected by the plasmid can grow. This eliminates all the cells in category 1. Alternatively, all the cells can be spread onto an agar plate that contains tetracycline. Then only the cells containing a plasmid having a gene for tetracycline resistance will grow and form colonies.

The second trick distinguishes cells containing plasmids joined to rabbit DNA from those that do NOT have rabbit DNA attached to them (Figure 8-2). To execute this trick one uses a cloning vehicle, a plasmid, with two genes for antibiotic resistance. Often one gene is for tetracycline resistance (*tet*^{R}) and the other for ampicillin (penicillin) resistance (*amp*^{R}). Since restriction endonucleases cut in very specific locations, an endonuclease can be found that cuts the plasmid DNA only inside the gene for ampicillin resistance (Figure 8-3). Consequently, whenever rabbit DNA is attached to this plasmid, it will be inserted into the middle of the ampicillin-resistance gene, for that is where the ends of the DNA occur. The large stretch of rabbit DNA inserted into this plasmid gene will render the gene inactive. Consequently, cells containing plasmids with rabbit DNA inserted into the ampicillin-resistance gene will be resistant only to tetracycline. On the other hand, cells containing a plasmid having no rabbit DNA will be resistant to both drugs. Thus all the colonies that formed on the tetracycline-containing agar plate (Figure 8-2*d*) must be tested to find ones that *fail* to grow on ampicillin-containing agar.

A piece of sterile velvet is carefully placed on the surface of the tetracycline-containing agar plate so it touches the bacterial colonies. Some of the cells from each colony will stick to the velvet, which is then removed and set onto a clean ampicillin-containing agar plate. Cells from each colony will come off the velvet and stick to the agar of the clean plate. There the cells will grow into colonies if they are resistant

---

Figure 8-2 Procedure for Obtaining Pure Clones Containing Rabbit DNA. (**a**) Plasmid DNA conferring resistance to tetracycline and ampicillin is mixed with rabbit cDNA. The DNAs are treated with a restriction endonuclease that cuts the plasmid only once, in the ampicillin-resistance gene. The endonuclease is then inactivated by heating the mixture to 60°C. (**b**) DNA ligase is added to join the DNA fragments. The rabbit cDNA becomes inserted into the middle of the ampicillin-resistance gene, inactivating it (see Figure 8-3). (**c**) *E. coli* cells are transformed with the recombinant DNA. (**d**) Plasmid-containing cells are selected by growth on agar containing tetracycline. Cells with plasmids joined to rabbit cDNA are identified by screening on ampicillin-containing agar (**e**) [these cells grow only on tetracycline (**f**)]. Pure cultures (**g**), which contain cloned genes, result.

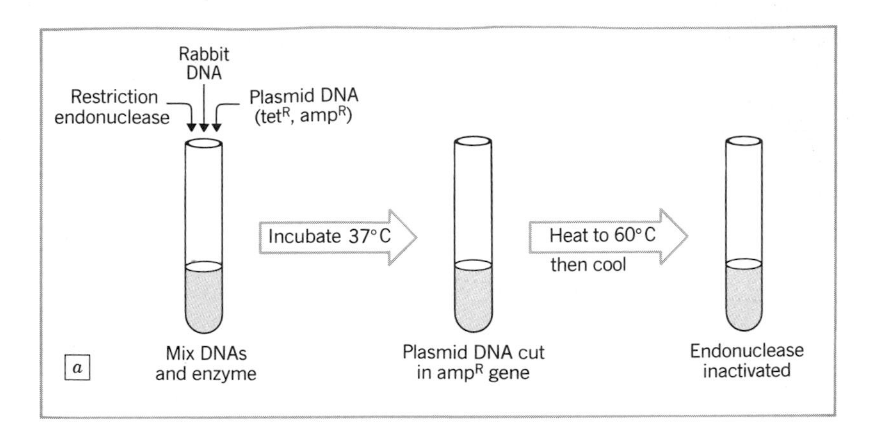
Rabbit DNA
Restriction endonuclease
Plasmid DNA (tet^R, amp^R)
Incubate 37°C
Heat to 60°C then cool
a
Mix DNAs and enzyme
Plasmid DNA cut in amp^R gene
Endonuclease inactivated

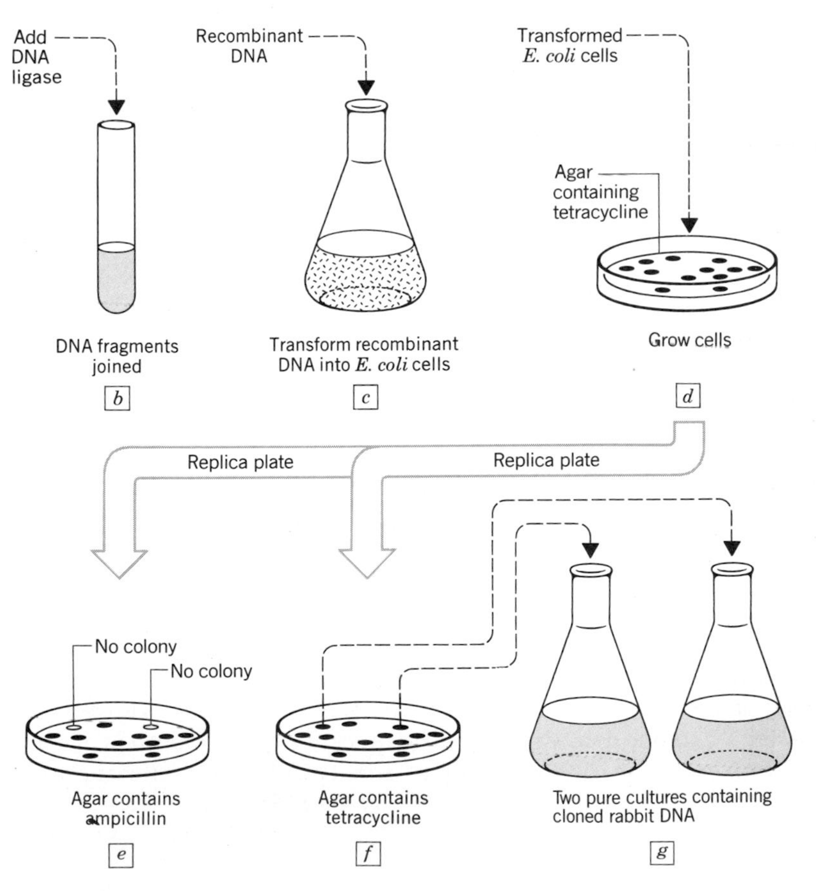
Add DNA ligase
Recombinant DNA
Transformed *E. coli* cells
Agar containing tetracycline
DNA fragments joined
b
Transform recombinant DNA into *E. coli* cells
c
Grow cells
d
Replica plate
Replica plate
No colony
No colony
Agar contains ampicillin
e
Agar contains tetracycline
f
Two pure cultures containing cloned rabbit DNA
g

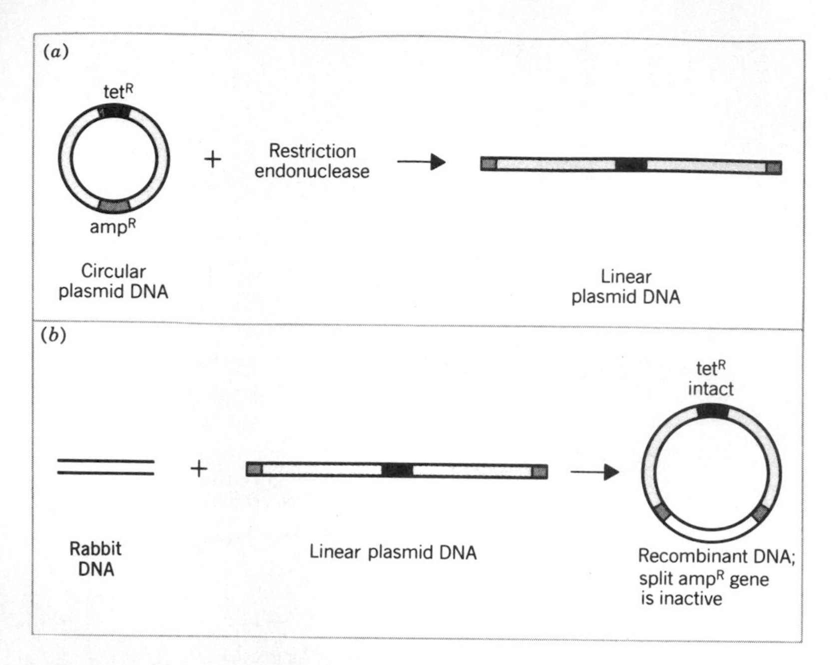

Figure 8-3 Inactivating a Gene. (**a**) Plasmid DNA containing genes for resistance to ampicillin and to tetracycline is cut with a restriction endonuclease, dividing the amp^R gene into two parts. (**b**) When rabbit DNA is inserted into the plasmid, the two parts of the amp^R gene remain separated; hence the gene is inactive.

to ampicillin. This technique is called **replica plating**, and it distributes cells onto the second plate in a pattern identical to the distribution of the colonies on the first plate. This procedure is outlined in Figure 8-2; two colonies are present in Figure 8-2*f* that are absent in Figure 8-2*e*. These colonies contain rabbit DNA inserted into the plasmid.

But at this stage one doesn't know which gene or genes are contained in any particular colony identified above. In fact, the chances of obtaining one that has the gene of interest can be very low. The replica plating technique makes it possible to quickly screen thousands of colonies for growth in different drugs by simply looking for differences in the distribution patterns of colonies on agar plates.

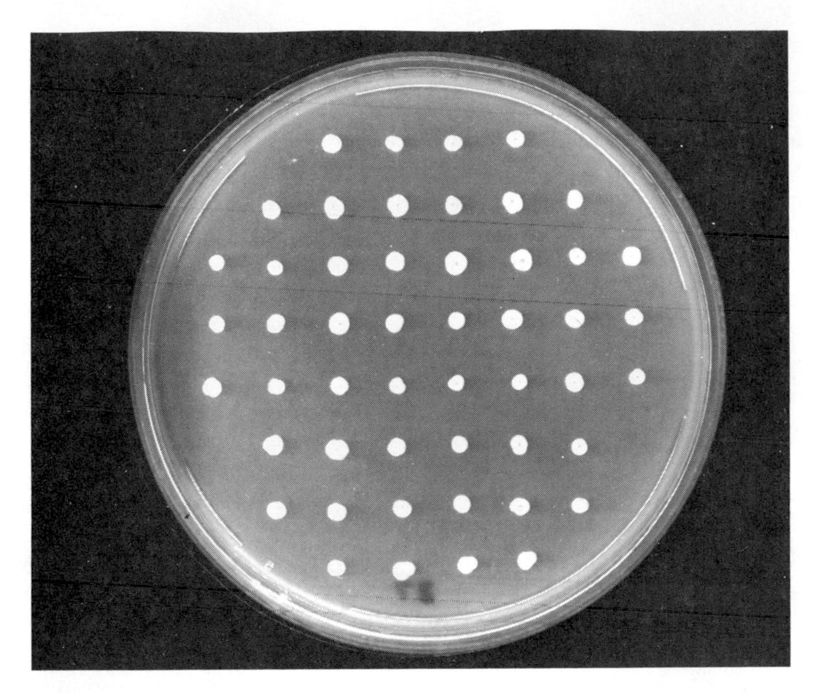

Figure 8-4 Agar Plate Bacterial Colonies Arranged in a Regular Grid Pattern.

The next trick is to identify bacterial colonies that have the particular gene being sought. First, a small sample of cells from each colony containing a rabbit DNA fragment (Figure 8-2*f*) is touched with a sterile toothpick and transferred to a new agar plate, where a gridlike pattern of colonies will be set up (Figure 8-4). Sometimes a cloned gene will give its bacterial host a selective growth advantage; in such cases the components of the agar are adjusted so that the only colonies to grow will be those that contain the gene being sought. This is rarely the case with genes from complex organisms such as rabbits or humans. Usually, the appropriate colony is located by nucleic acid hybridization or by reaction of the colony with antibodies specific to the protein product of the gene being sought. Hybridization is discussed in the next section.

SCREENING BY NUCLEIC ACID HYBRIDIZATION

Screening bacterial colonies for a particular gene is carried out in the following way. First, colonies are grown on an agar plate. Then a piece of filter paper is placed on the agar and removed. Bacterial cells stick to the paper, forming the same grid pattern they had held on the agar (sometimes the cells are grown directly on filter paper placed on the agar surface). The paper is then transferred to a dilute solution of **sodium hydroxide** (lye or caustic soda). The sodium hydroxide breaks the cells, and some of the cellular debris plus the cellular DNA stick tightly to the paper. The sodium hydroxide also causes the DNA to become single-stranded. Since the bacterial colonies had been arranged in a grid on the paper, DNA released from the cells forms an identical pattern. Next, the sodium hydroxide is neutralized with acid, and the paper is slipped into a dish containing a radioactive probe. This probe is often made synthetically from information derived from the amino acid sequence of the protein.

Both the radioactive probe DNA and the cellular DNA attached to the paper are single-stranded, ready to form base pairs with any DNA they contact. But the probe will form base pairs with paper-bound DNA only if the paper-bound DNA contains the gene being sought (see Figure 5-10). When this happens, the radioactive DNA will be indirectly bound to the paper, and the location of the radioactivity will identify the bacterial colony that contains the rabbit gene being sought.

To determine where the radioactivity is located, the paper is removed from the dish and washed thoroughly to remove any radioactive probe that is not base-paired with paper-bound DNA. X-ray film is then placed next to the paper (Figure 8-5*a*). Wherever radioactive probe is base-paired to paper-bound DNA, it exposes the film, producing a dark spot (Figure 8-5*b*). These dark spots correspond to bacterial colonies containing the gene of interest. This searching procedure makes it possible to obtain a pure culture of *E. coli* in which each cell contains a plasmid onto which the gene of interest has been cloned.

Screening the lambda plaques for those containing rabbit hemoglobin genes was carried out in a similar way. Hemoglobin cDNA was

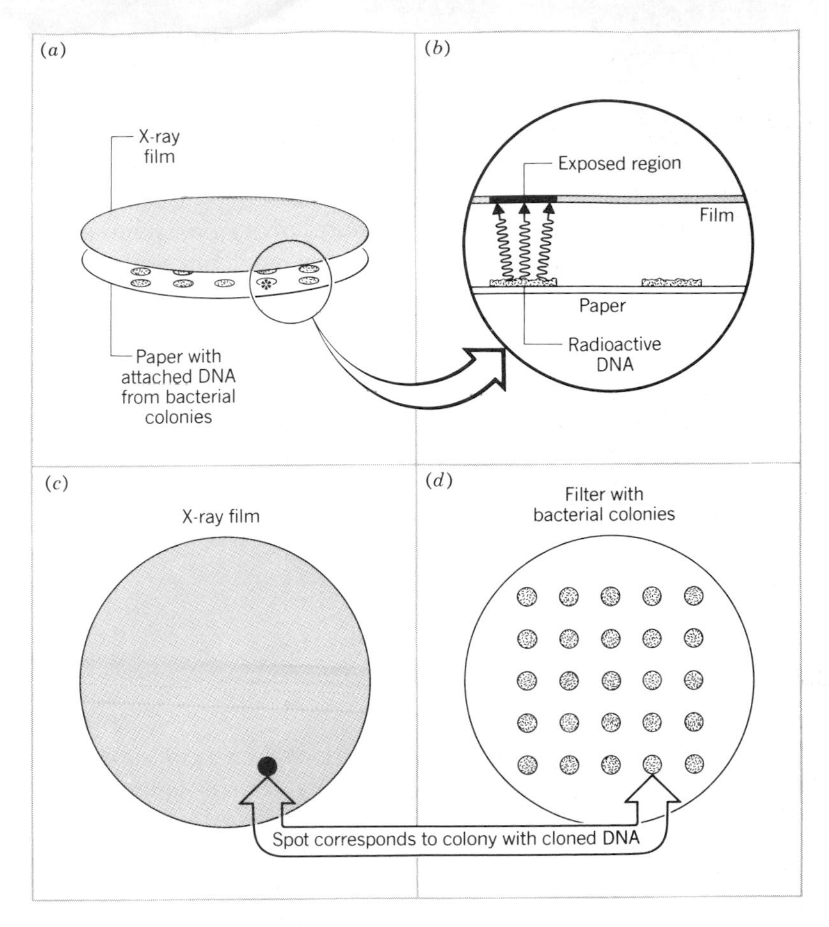

Figure 8-5 Detection of Bacterial Colonies Containing a Particular Rabbit Gene. Bacterial cultures are grown on paper placed on the surface of an agar plate (nutrients seep through the paper). The paper is removed and is dipped in sodium hydroxide to break open the cells in the colonies and denature the DNA. The denatured (single-stranded) DNA becomes attached to the paper. The sodium hydroxide is neutralized, and radioactive probe DNA is added. Complementary base pairs form between the probe DNA and paper-attached DNA if the gene being sought is present in the bacterial colony (see Figure 5-10). The paper is washed and placed next to X-ray film (**a**). If a particular bacterial colony contains the gene of interest [* in (a)], its DNA will have hybridized to the radioactive probe, and the X-ray film will be exposed above the colony (**b**). The pattern of exposed spots on the film (**c**) is used to identify the cultures containing cloned genes by comparison with the original distribution of colonies on filter paper (**d**).

cloned into a plasmid, and large amounts were obtained by purification from bacterial cells. Next, the cDNA plasmids were linearized by cutting with a restriction endonuclease and radioactively labeled using an enzyme that could add a radioactive phosphorus atom to the end of DNA. This generated a radioactive probe for hemoglobin genes. More than 750,000 phage plaques were examined, leading to the discovery of four containing hemoglobin genes. These clones, obtained directly from rabbit DNA, are called **genomic clones** to distinguish them from the cDNA clones derived from mRNA. This distinction is important, for as described in Chapter 10, mRNA from higher organisms often contains only parts of the nucleotide sequence of the gene. Thus cDNA clones would also contain only parts of the gene.

PERSPECTIVE

Often existing technology dictates the direction taken by research. For example, in the mid-1970s methods were available for isolating hemoglobin mRNA from red blood cells, the cells specialized to make hemoglobin. These mRNA molecules could be used as templates for the synthesis of radioactive probes to locate cloned hemoglobin genes. This is a rare situation, for most other genes, which might have been just as interesting to study as hemoglobin, are transcribed into RNA in cells that make mRNA of so many different types that they are not easily separated from each other to make hybridization probes for particular genes. Thus hemoglobin genes were an obvious choice for these early cloning studies.

The cloning strategies themselves also illustrate how biological research builds on previous developments. Messenger RNA isolation, phage and plasmid manipulation, restriction endonuclease cutting of DNA, enzymatic synthesis of DNA, and nucleic acid hybridization were all highly refined technologies being used for other studies before they were used for gene cloning. Gene cloners combined these methodologies in a new way to serve a new purpose. Indeed, most progress in science comes from combining selected aspects of existing knowledge.

Questions for Discussion

1. What information do you need to have about a gene to clone it?
2. Suppose that you wish to clone the gene that codes for tetracycline resistance. Devise a cloning procedure that would require no radioactive probes.
3. If the amino acid sequence of a protein is known, it is sometimes possible to use the genetic code (Figure 3-6) to help clone the gene. This is done by deducing the nucleotide sequence for a short region of the gene from the amino acid sequence of the protein. Next an oligonucleotide (usually about 15 nucleotides long) is synthesized which has the nucleotide sequence predicted from the amino acid sequence of the protein. This oligonucleotide is then radioactively labeled and used as a probe for detecting bacterial colonies or phage plaques that contain the gene. The most suitable oligonucleotide would be constructed from a region of the protein that would be rich in which amino acids? Regions rich in which amino acids would be least suitable?
4. Suppose that you know the nucleotide sequence of a particular gene in chimpanzees and wish to clone the homologous gene from human DNA. How would you use the polymerase chain reaction (Chapter 5) to help you clone the gene? Some chimpanzee and human genes differ by fewer than one nucleotide out of a hundred thousand. Is this relatedness close enough to permit you to expect success using the polymerase chain reaction?
5. The polymerase chain reaction is a way to amplify specific regions of DNA for subsequent nucleotide sequence analysis. Thus it is very useful for genetic screening when the nucleotide sequence of a gene is already known. But can the polymerase chain reaction also be used to clone genes if only the amino acid sequence of the protein product of the gene is known? If so, how? If not, why not?
6. Suppose you wish to clone a gene encoding a protein whose sequence is not known, but you do have antibodies that will bind specifically to the protein and cause it to precipitate (if you are unfamiliar with antibodies, see Chapter 10 and Appendix III). How could the antibodies be used to obtain the messenger RNA from the gene encoding the protein? (Hint: see frontispiece of Chapter 4).

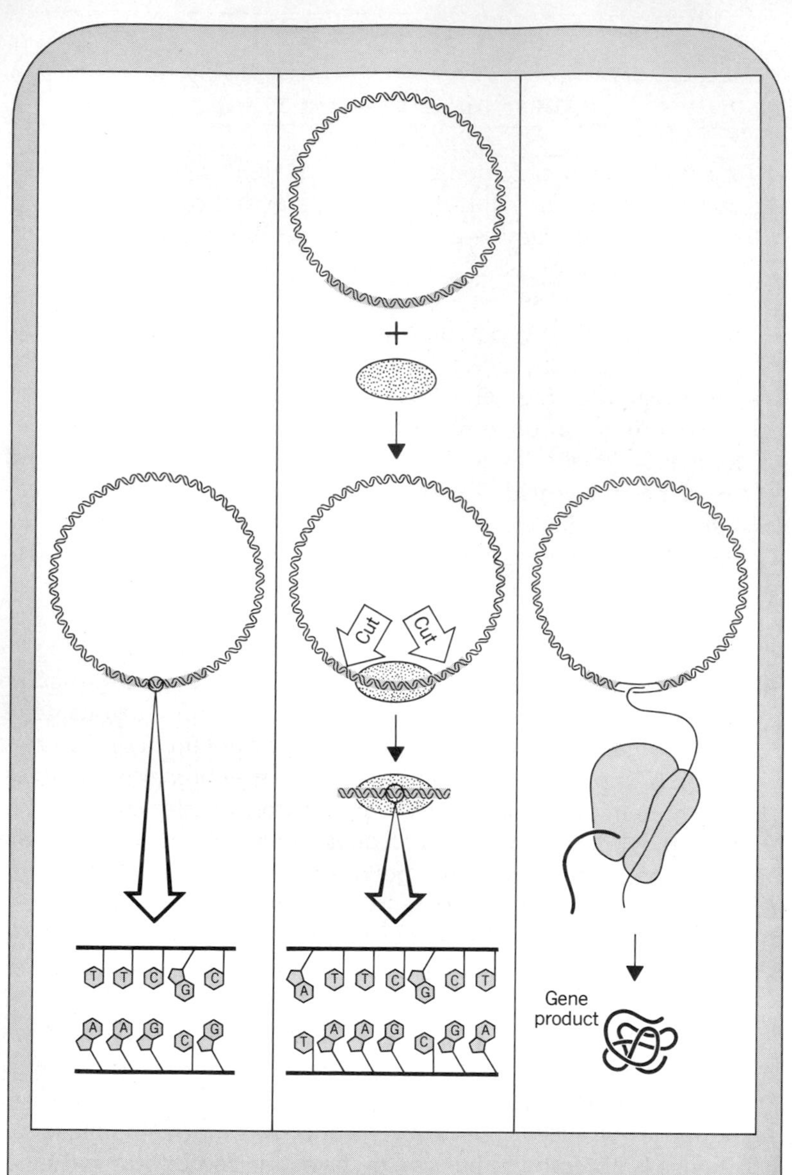

Cloned genes are used to determine nucleotide sequences, to identify sites where proteins bind to DNA, and to produce large amounts of gene product.

CHAPTER NINE

CLONED GENES AS RESEARCH TOOLS

Gene Structure, Expression Vectors, Southern Hybridization, and Binding Proteins to DNA

Overview

Gene cloning technologies make it possible to obtain the large amounts of specific regions of DNA necessary to study gene structure and function. Gene structure is analyzed by determining the nucleotide sequence of a section of DNA. In one method a short region of DNA is first purified and cloned into a bacterial virus called M13. DNA polymerase is used to produce a series of DNA fragments that extend from a specific spot in M13 DNA to particular nucleotides in the sequence being analyzed. By measuring the lengths of these fragments, it is possible to determine the position of each nucleotide relative to the fixed spot in M13 DNA. The nucleotide sequence can then be deduced. The exact position of a gene can be located by finding a nucleotide sequence identical to that predicted from the amino acid sequence of the protein product of the gene.

Changes in gene structure and location can be detected even without determining the nucleotide sequence, for often such changes alter the restriction map of the gene and the surrounding regions. These alterations can be seen by nucleic acid hybridization methods using cloned probes.

Gene function is studied by examining how the protein product of a gene interacts with other proteins, with DNA, and with small molecules. Gene cloning makes it possible to produce the large amounts of particular proteins and particular regions of DNA needed for this type of study.

INTRODUCTION

In the first eight chapters we focused on the concepts of molecular biology to explain the process of gene cloning. It may have appeared that cloning a particular gene is an end unto itself. It is not. Gene cloning is but a tool, albeit a very powerful tool, that biologists use to explore life. In this chapter we examine several research applications of gene cloning.

Gene cloning is a way to obtain large amounts of short, specific regions of DNA. These sections of DNA that are dissected out can be very short, often representing less than one-millionth (0.000001) of the total nucleotide sequence of an organism's DNA. By using gene cloning we can obtain enough DNA and protein molecules to study gene structure, function, and regulation with remarkable precision.

ANALYSIS OF GENE STRUCTURE

The initial step in analyzing gene structure is to determine the nucleotide sequence of the gene. DNA molecules are incredibly long and monotonous; deciphering the exact order of thousands of A's, T's, G's, and C's requires an ingenious combination of many of the molecular tools discussed in earlier chapters.

The recipe outlined in Chapter 8 makes it possible to insert a particular DNA fragment into either plasmid DNA or phage DNA. The resulting recombinant DNA is copied trillions of times by bacteria, and these molecules are purified. Further analysis sometimes requires that the cloned DNA be separated from the cloning vehicle DNA (the plasmid or phage DNA). Restriction endonucleases and gel electrophoresis can be used to accomplish this. Often the cloned DNA fragment is inserted into the cloning vehicle at a site where a specific restriction endonuclease cuts; if the joining event regenerates the recognition site for the enzyme (Figure 9-1), the recombinant DNA need only be treated with the same restriction endonuclease to liberate the DNA fragment from the cloning vehicle (see Figure 7-6). Since the cloning vehicle and the cloned DNA fragment usually have different lengths, they can be physically separated into discrete bands by gel electrophoresis (see Figures 7-4 and 9-2). DNA from a particular band can be recovered by slicing the gel where the band is located, transferring the

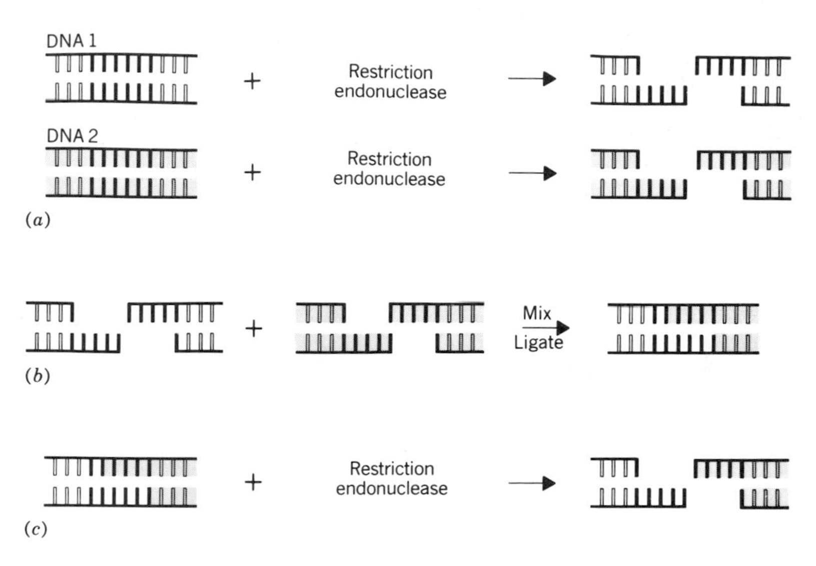

Figure 9-1 Regeneration of a Restriction Site. **(a)** Two DNA molecules containing a recognition site for the same restriction endonuclease (solid regions) are cut with that enzyme. When the DNA molecules are ligated together **(b)**, the restriction site is sometimes regenerated. The ligated DNA can in turn be recut into fragments **(c)** by subsequent treatment with the restriction endonuclease. Plasmids have been constructed which have special sites where many different restriction endonucleases cut. Thus a variety of restriction endonucleases can be used to excise the fragment.

slice to a second electrophoresis apparatus, and then driving the DNA out of the gel slice by applying an electric current. If necessary, the DNA fragments can be cut into still smaller fragments with other restriction endonucleases, and these new fragments can be further separated by gel electrophoresis or recloned by insertion into a plasmid or phage DNA. Thus the DNA fragments can be gradually pared down to obtain pieces small enough for nucleotide sequence analysis.

Although several methods have been developed to determine nucleotide sequences, all use the same general strategy: a reference point is established, and the distance is measured from the reference point to each of the nucleotides of a given type—for example, to each A, to

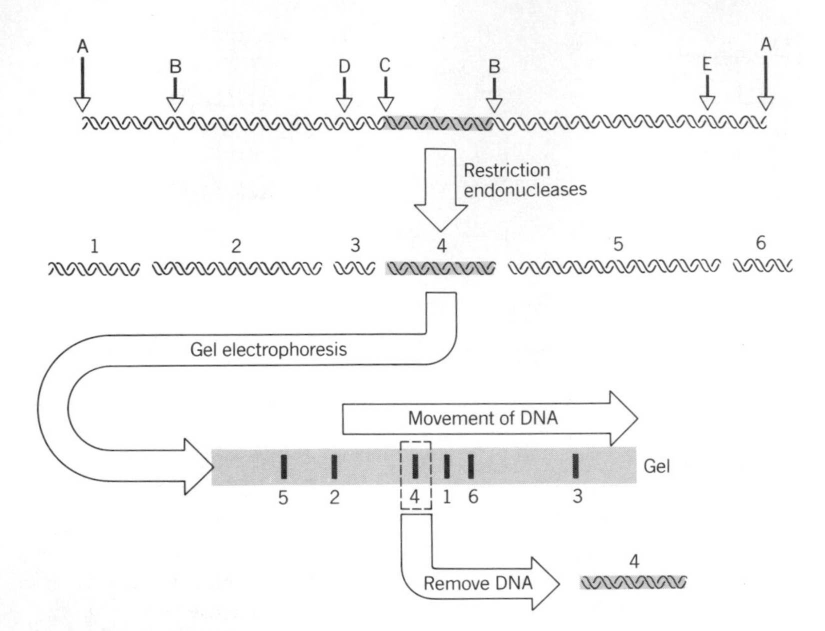

Figure 9-2 Cutting DNA into Pieces of Manageable Size. A cloned human DNA fragment is often much longer than the region being sought (shaded region); thus, it must be cut into smaller pieces. Arrows **A** represent the ends of the human DNA fragment, and arrows **B**, **C**, **D**, and **E** indicate cleavage sites for four different restriction endonucleases. Treatment of the human DNA fragment with these four nucleases produces fragments 1 through 6; these fragments are physically separated by gel electrophoresis. Each band in the gel (Figure 7-4) results from billions of identical DNA molecules. Individual bands are cut out of the gel, and the DNA is removed. In this example, band **4** would be saved for further study.

each G, to each C, or to each T. In one method, the DNA fragment under study is first cloned into a specific site in the DNA of a bacteriophage called M13. During infection of bacteria, this particular phage packages only one of its DNA strands; thus, recombinant DNA isolated from these virus particles will contain only one of the two DNA strands (see Figure 9-3). To establish a reference point, the DNA is mixed with short, single-stranded pieces of DNA that are complementary to a region of M13 DNA near the position where the gene to be sequenced has been inserted. The short pieces form base pairs with the

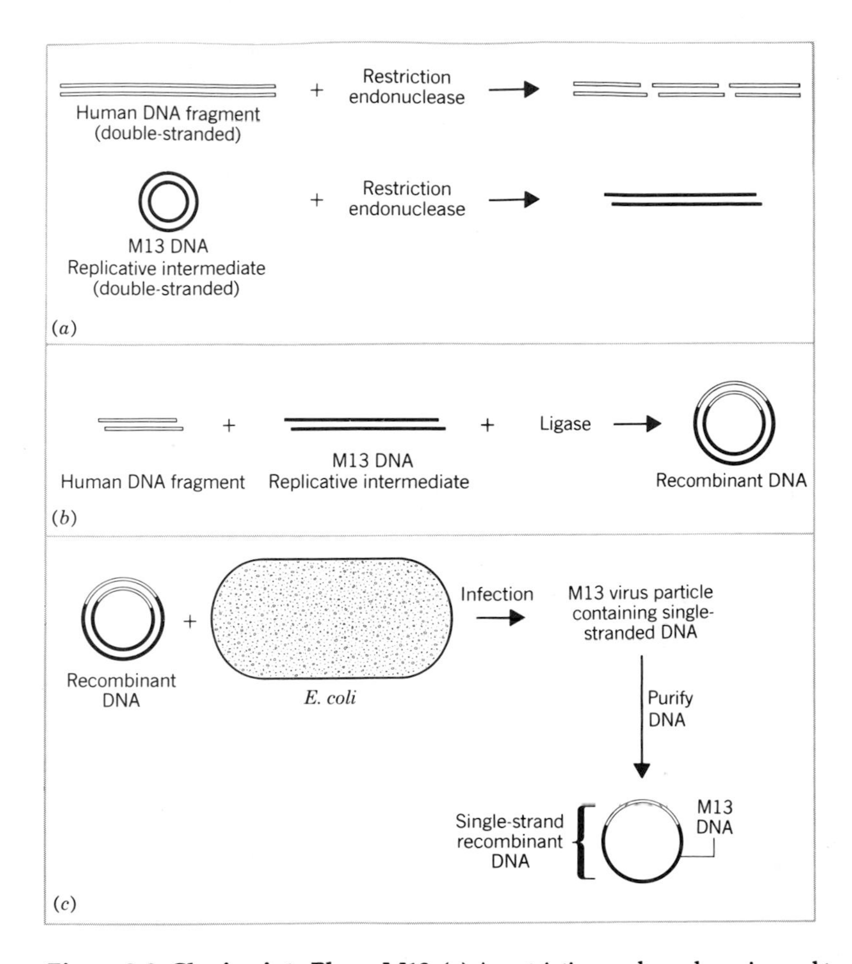

Figure 9-3 Cloning into Phage M13. (a) A restriction endonuclease is used to cut a human DNA fragment and a purified double-stranded replicative intermediate of phage M13 (single-stranded DNA viruses form double-stranded DNA molecules as a part of their life cycle). **(b)** The DNAs are mixed and ligated to form a recombinant DNA. **(c)** The recombinant DNA is used to infect *E. coli* cells, producing M13 virus. The virus particles contain only one of the two DNA strands. This DNA strand, which is the same for all the virus particles, can then be purified in large quantities.

M13 DNA, creating a short, double-stranded region of DNA that can serve as a primer for synthesis (Figure 9-4*a*). A large number of these partially double-stranded DNA molecules are mixed with DNA polymerase and radioactive nucleotides. DNA polymerase synthesizes radioactive DNA from the M13 template, beginning at one end of the short, double-stranded region (DNA polymerase requires a primer to start synthesis of DNA; thus, its synthesis begins only where the primer is located). DNA polymerase soon crosses into the cloned DNA region and uses it as a template to make new DNA. The polymerase can be stopped by including in the reaction a nucleotide analogue (a fake nucleotide) that lacks a 3′ oxygen (see Figure 3-2). Such a nucleotide cannot be part of a DNA backbone and will not allow the new DNA chain to grow. If the analogue behaves like an A, the new, radioactive chain will stop after it would normally have added an A (Figure 9-4*b*). Normal A's are included in the reaction mixture to allow some DNA synthesis before an A analogue stops the reaction. Since there are many A's in a nucleotide sequence, the stops will not always be in the same place. Thus, by carefully adjusting the amount of analogue in the reaction mixture, it is possible to create a collection of radioactive DNAs that begin at a distinct spot (the end of the primer) and stop at the various positions where A's occur. The identical procedure is repeated in three other test tubes using the respective analogues for G, C, and T. Thus four separate collections of molecules are made. Every molecule starts at the same place, and the various types end at different distances from the starting point. Figure 9-5 illustrates such a collection in which the reaction was stopped by an analogue of A.

Figure 9-4 Creating a Collection of Variable Length Copies of Cloned DNA. (**a**) Many single-stranded M13 recombinant DNA molecules from Figure 9-3 are mixed with short single-stranded DNA fragments complementary to a region of M13 DNA near the junction between M13 DNA and the cloned DNA. The fragment forms a double-stranded region with M13 DNA. (**b**) The short double-stranded region acts as a primer for DNA polymerase. New DNA is synthesized from radioactive A's, T's, G's, and C's onto one end of the primer. An analogue of A (A*) is added to halt synthesis at the various positions where its complement, T, occurs in the nucleotide sequence of the cloned DNA. (**c**) The DNAs are treated so that all of them become single-stranded. Their sizes are then measured by gel electrophoresis.

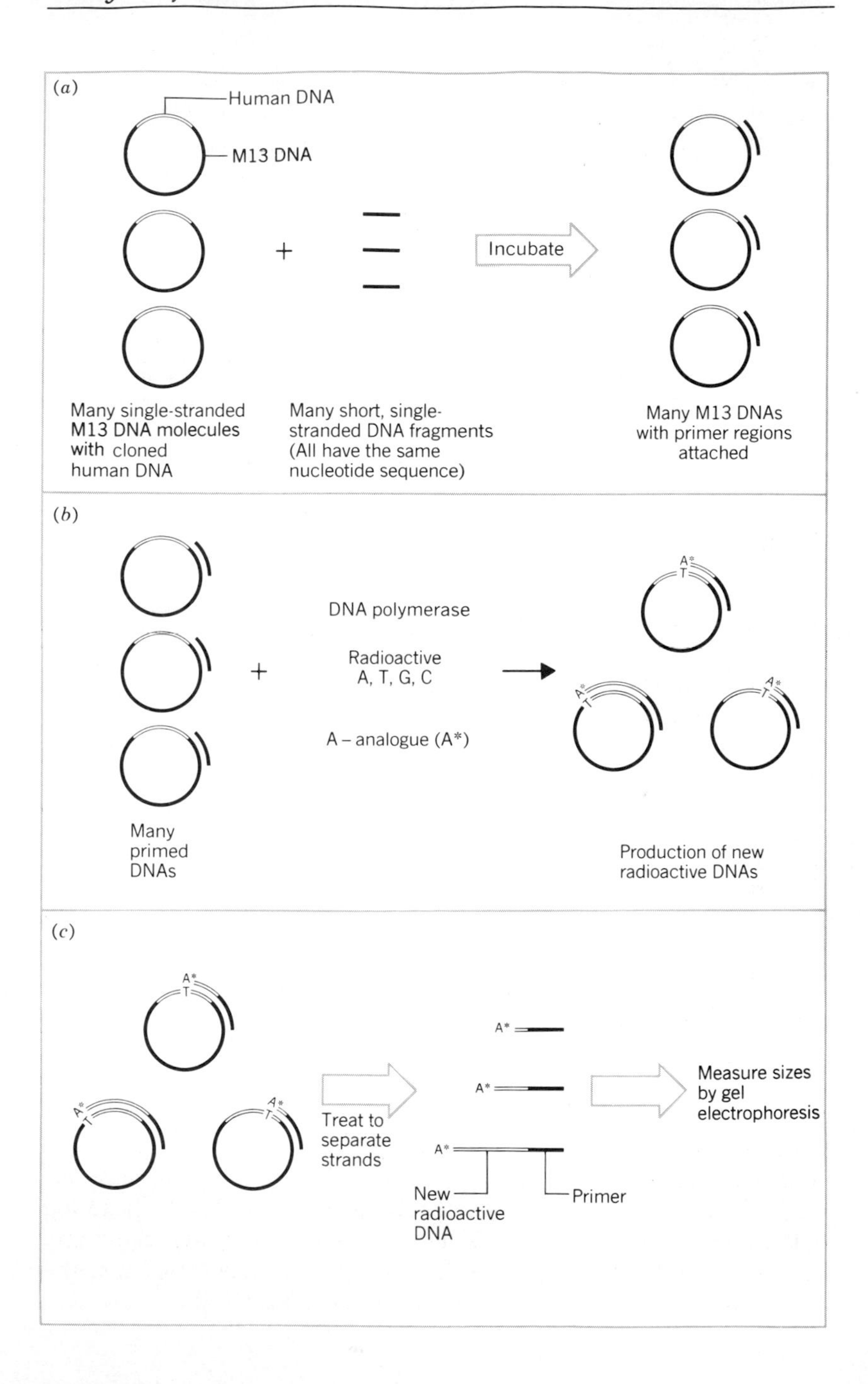
(a)
Human DNA
M13 DNA
+
Incubate
Many single-stranded M13 DNA molecules with cloned human DNA
Many short, single-stranded DNA fragments (All have the same nucleotide sequence)
Many M13 DNAs with primer regions attached
(b)
DNA polymerase
Radioactive A, T, G, C
A – analogue (A*)
+
Many primed DNAs
Production of new radioactive DNAs
(c)
Treat to separate strands
Measure sizes by gel electrophoresis
New radioactive DNA
Primer

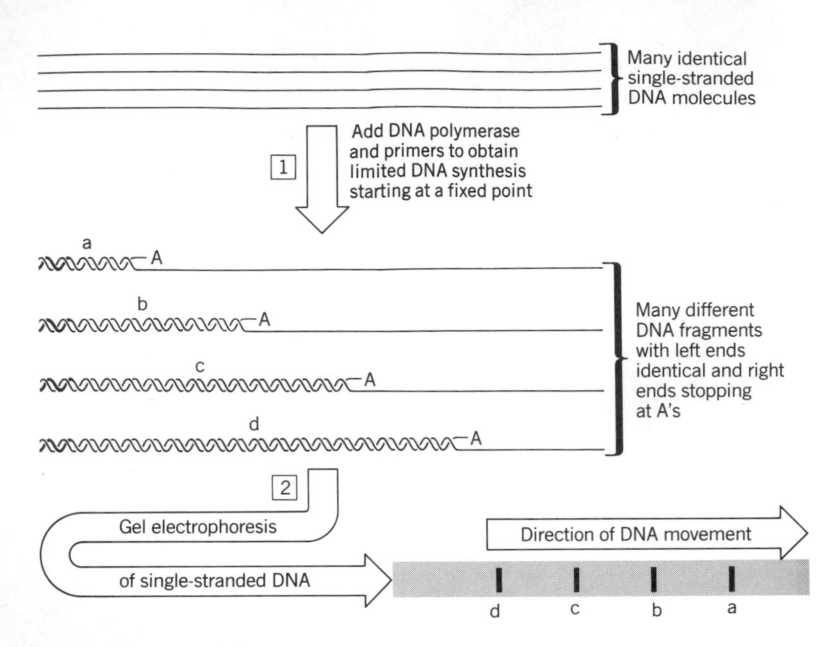

Figure 9-5 Creating and Analyzing a Collection of DNA Fragments Having Different Lengths. (**1**) Many identical single-stranded DNA molecules are partially replicated. A collection of shorter molecules (**a–d**) is formed. All the new DNAs have the same left end but different right ends, which in this example always stop at an **A**. (**2**) The collection of fragments is denatured (converted into single strands) and analyzed by gel electrophoresis to determine the length from the left end to the position of each **A** (the distance a DNA molecule moves in the gel during electrophoresis is related to its length).

The lengths of the newly synthesized molecules are measured by gel electrophoresis under conditions in which the DNA molecules are in a single-stranded configuration (Figure 9-5). The electrophoresis methods are precise enough to permit the separation of DNA molecules differing by only one nucleotide. The four collections of molecules (terminated at either A's, G's, C's, or T's) are subjected to electrophoresis in lanes next to each other, and after electrophoresis a piece of film is exposed by the radioactivity in the DNA. By examining only radioactive molecules, it is not necessary to consider the many nonradioactive DNA fragments that might otherwise complicate the experiments. A series of bands is seen (Figure 9-6). The lowest band in

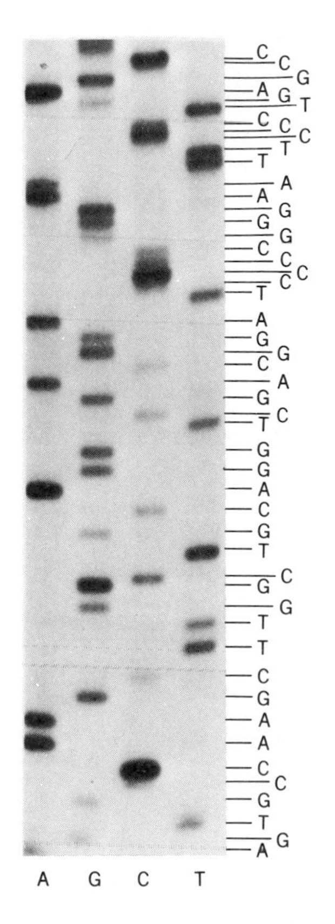

Figure 9-6 DNA Sequencing Gel. Four radioactive DNA samples were electrophoresed in adjacent lanes. The four samples labeled A, T, C, and G across the bottom of the figure were synthesized in the presence of analogues to A, T, C, and G, respectively, as described in Figure 9-4. The nucleotide sequence of the DNA is indicated on the right. (Photograph courtesy of Richard Archer, Janice Zengel, and Lasse Lindahl, University of Rochester.)

Figure 9-6 represents molecules that extend to an A, for that band appears in the sample containing the A analogue (A*). The next higher band is in the G-analogue lane, so the next nucleotide is a G. Thus nucleotide sequences can be read directly from pictures such as that shown in Figure 9-6.

The nucleotide sequences from a number of adjacent restriction fragments can be fit together to produce very long sequences. For example, the entire sequence for bacteriophage lambda DNA (48,498 nucleotides) was determined in this way. Particular genes are located in the sequence by comparing the nucleotide sequence to that expected from the amino acid sequence of the protein made from the gene. For example, if the left-most amino acid in the protein is methionine, the second one from the left is tryptophan, and the third one is phenylalanine, we expect the nucleotide sequence of that portion of the gene to be A T G T G G T T (T or C) because we know the triplet codon for each amino acid (see Figure 3-6). The ninth nucleotide could be either T or C because phenylalanine is encoded by two triplets, T T T and T T C. This type of comparison allows us to determine the exact position of a gene. Nucleotide sequences can also be scanned for regions capable of encoding a protein. These regions, called open reading frames (ORFs), contain long stretches of nucleotides between start and stop codons, and they frequently signify the location of previously unidentified genes. Control regions can also be spotted when their sequences are similar to those known to control the expression of other genes. Thus nucleotide sequence analysis is a major tool in studying the anatomy of DNA.

EXPRESSION VECTORS

Biologists often wish to obtain large amounts of protein from the gene they have cloned, and one way to accomplish this is to move the cloned gene into a specially constructed plasmid called an **expression vector** (see Figure 9-7). The expression vector contains an easily controlled promoter such as that from the *lac* operon (see Chapter 4) situated upstream from a series of sites for restriction endonucleases. Thus the expression vector plasmid can be easily cut by a variety of nucleases immediately downstream from the promoter. Once the general features of the cloned gene (i.e., the location of the beginning and end of the gene) have been determined, the cloned gene can be trimmed and tailored to permit insertion into the expression vector next to the promoter. Bacterial cells are then transformed with the expression vector, and when the inducer of the promoter is added to the cell culture, the bacterial cells will begin to produce protein from the cloned gene. Systems have been developed in which 40% of the cellular pro-

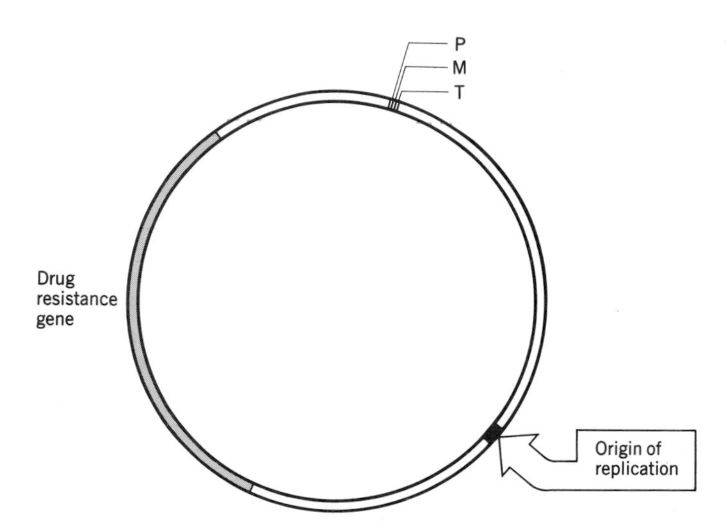

Figure 9-7 A Plasmid Expression Vector. In this circular map of a plasmid, an inducible promoter, labeled **P**, is followed closely by a multiple cloning site **M** (a short region of about 40 base pairs constructed to allow many different restriction endonucleases to cleave the DNA there). Farther downstream is a DNA sequence that serves as a terminator (**T**) to stop transcription. A gene of choice is placed in the plasmid at the multicloning site. The plasmid also contains an origin of replication (to allow it to multiply inside bacteria) and a drug-resistance marker (to enable the biologist to select for bacteria that harbor the plasmid). Promoters are used which are controlled by a repressor that can be easily regulated by additions to the growth media of the cells; thus production of the protein of choice can be induced whenever desired.

tein is made from the cloned gene. Expression vectors that function in higher cells are also available.

SOUTHERN HYBRIDIZATION

Very small alterations in the nucleotide sequence of DNA molecules can be detected by analysis of changes in location of cuts by restriction endonucleases. These changes alter the size of the DNA fragments produced and thus the restriction maps of the DNA molecules (see Figure 7-5 for a hypothetical example). Since large DNA molecules have very complicated restriction maps, a method has been devised to examine only a small, specific region of the entire DNA. In this method, called **Southern hybridization** (or Southern blotting), a radioactive probe, complementary to the region of interest, is used to identify specific DNA fragments by nucleic acid hybridization. The procedure is outlined in Figure 9-8. After DNA has been cut and the fragments separated from one another by gel electrophoresis, the fragments are denatured and transferred from the gel to nitrocellulose paper. There they can be hybridized to a radioactive probe that contains nucleotide sequences from the gene or region of interest (see Figure 5-10). After incubation with the probe, nonhybridized radioactivity is washed off the paper, and the paper is placed next to a piece of X-ray film to expose it. Since the relative position of each fragment is maintained during the transfer from gel to paper, the position of the radioactive band can be used to calculate the size of the fragment containing nucleotide sequences identical to those in the probe.

Southern hybridization, combined with cleavage of DNA by restriction endonucleases, has become a popular tool for **prenatal** diagnosis of certain genetic defects. In Chapter 7 it was pointed out that genetic diseases are often associated with changes in restriction maps. Southern hybridization makes it possible to limit the analysis to the region at or near the gene responsible for the disease. For such analyses a portion of the polymorphic region must be cloned to use as a probe.

Southern hybridization can also be used to detect changes in the lo-

---

Figure 9-8 Southern Hybridization. (**a**) Chromosomal DNA is cut with a restriction endonuclease to produce small DNA fragments. The gene of interest is shown as a solid region and in this example is found on only one type of fragment. The fragments are separated by gel electrophoresis. Many bands (fragments) are seen following staining of DNA (band labeled **B** contains the gene of interest). The DNA in the gel is denatured and transferred to nitrocellulose paper. (**b**) Cloned DNA containing the gene of interest is purified and cut with a restriction endonuclease to release a small fragment from within the gene of interest. This small fragment is purified by gel electrophoresis, radioactively labeled, and denatured. This DNA is called the probe. (**c**) The filter-bound DNA and the probe are mixed and incubated. After band **B** has hybridized to the probe and become radioactive, the nonhybridized radioactive DNA is removed. Band **B** is easily identified on the film. Its position on the paper is related to its length (smaller DNA will be nearer the bottom).

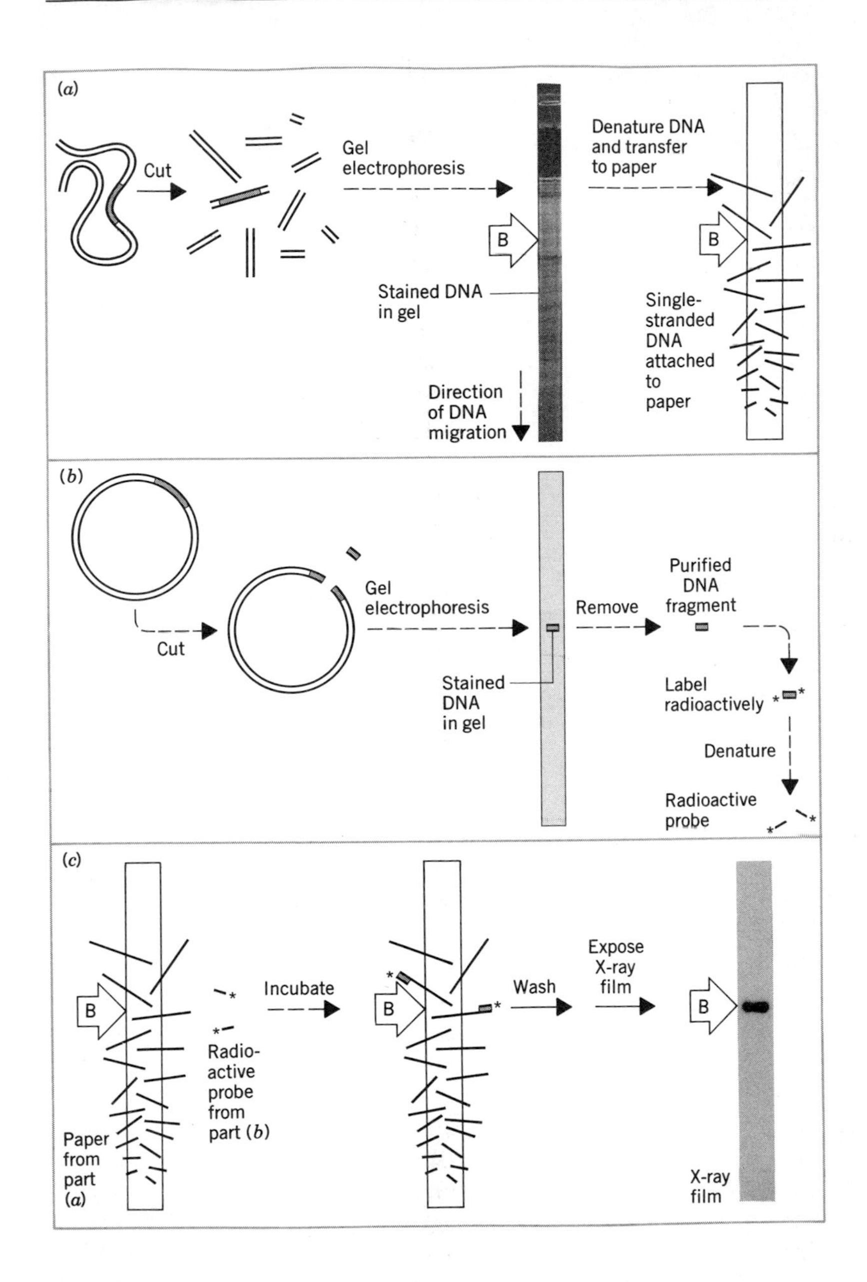
(a)
Cut
Gel electrophoresis
B
Stained DNA in gel
Direction of DNA migration
Denature DNA and transfer to paper
B
Single-stranded DNA attached to paper
(b)
Cut
Gel electrophoresis
Stained DNA in gel
Remove
Purified DNA fragment
Label radioactively
Denature
Radioactive probe
(c)
B
Paper from part (a)
Incubate
Radio-active probe from part (b)
B
Wash
Expose X-ray film
B
X-ray film

cation of specific regions of DNA. For example, the method was used to show that two strains of the same bacterial species contain *tst*, the gene for toxic shock syndrome, at different locations in their chromosomes (Figure 9-9). Could this mean that the *tst* gene is able to jump around in the DNA?

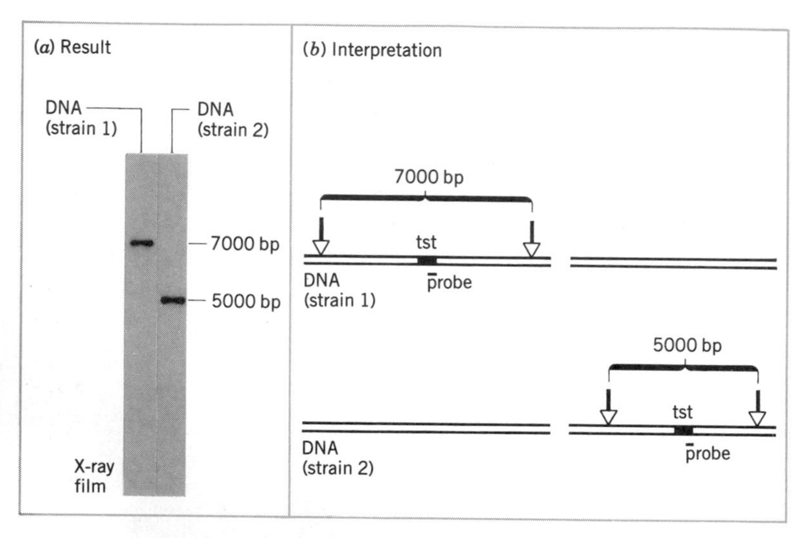

Figure 9-9 Location of Toxic Shock Syndrome Gene Varies Among Strains. DNA isolated from two strains of the bacterium *Staphylococcus aureus*, the organism that causes toxic shock syndrome, was examined as described in Figure 9-8. The DNA samples were digested with the restriction enzyme called ClaI, an enzyme that does not cut inside the toxic shock syndrome toxin-1 gene *tst*. The fragments were next electrophoresed in an agarose gel. The gel was treated with alkali to denature the DNA, and then it was placed on nitrocellulose paper to transfer the DNA fragments by a blotting technique. Following the transfer, the paper was incubated with a radioactive probe obtained from the cloned *tst* gene. After rinsing, the paper was used to expose X-ray film, and single bands of DNA were seen for each DNA sample. As seen in (**a**), one band contained 7000 base pairs (bp), the other 5000. For this to happen, the DNA surrounding the *tst* gene had to be different in the two strains; that is, the gene had to be at different positions in the chromosomes of the two strains (arrows in (**b**) indicate the locations of restriction endonuclease cleavage sites). Perhaps *tst* is associated with a transposon (see Chapter 10) that can move from one spot on the chromosome to another. (Photo courtesy of Barry Kreiswirth and Richard Novick, Public Health Research Institute.)

ANALYSIS OF GENE FUNCTION

The protein products of many important genes are present in very tiny amounts inside living cells, and it is often very difficult to obtain enough of a particular protein to study its properties and interactions with other molecules. But in a number of cases gene cloning and the use of expression vectors has erased the problem.

Some of these protein products are particularly interesting because they act on DNA. Repressors are an example. Chapter 4 pointed out that repressors prevent RNA polymerase from transcribing RNA from certain genes, and thus repressors regulate gene expression. Understanding this aspect of gene control involves knowing exactly where the repressor binds to the DNA. Gene cloning technologies have made it possible to obtain large amounts of both repressor protein and the potential DNA binding sites. When the two are mixed together, complexes form between the protein and the DNA. DNA in these complexes is protected from cleavage produced when nucleases are added to the preparation (Figure 9-10). Consequently, analysis of nucleotide sequences that survive nuclease treatment provides insight into repressor binding sites, and this information adds to our general understanding of gene organization and regulation (see Figure 4-7).

Since cells can be manipulated to produce large quantities of certain proteins, biologists have also asked questions about what happens to the expression of related genes. In bacteria, for example, there are many related genes encoding proteins that serve as components of ribosomes. Massive overproduction of one of these proteins shuts off the production of several others. Biologists are studying this regulatory process to determine how cells control the production of their many components.

PERSPECTIVE

We know in general terms how DNA acts as a repository for information. By determining the nucleotide sequences of DNA molecules, we are rapidly learning exactly what that information is. But this new knowledge will not suddenly tell us how life works. Fundamental problems remain to be solved even with organisms as simple as bacteria. For example, knowing the exact order of all 4,400,000 base pairs in *E. coli* DNA will not tell us how the timing of cell division is controlled.

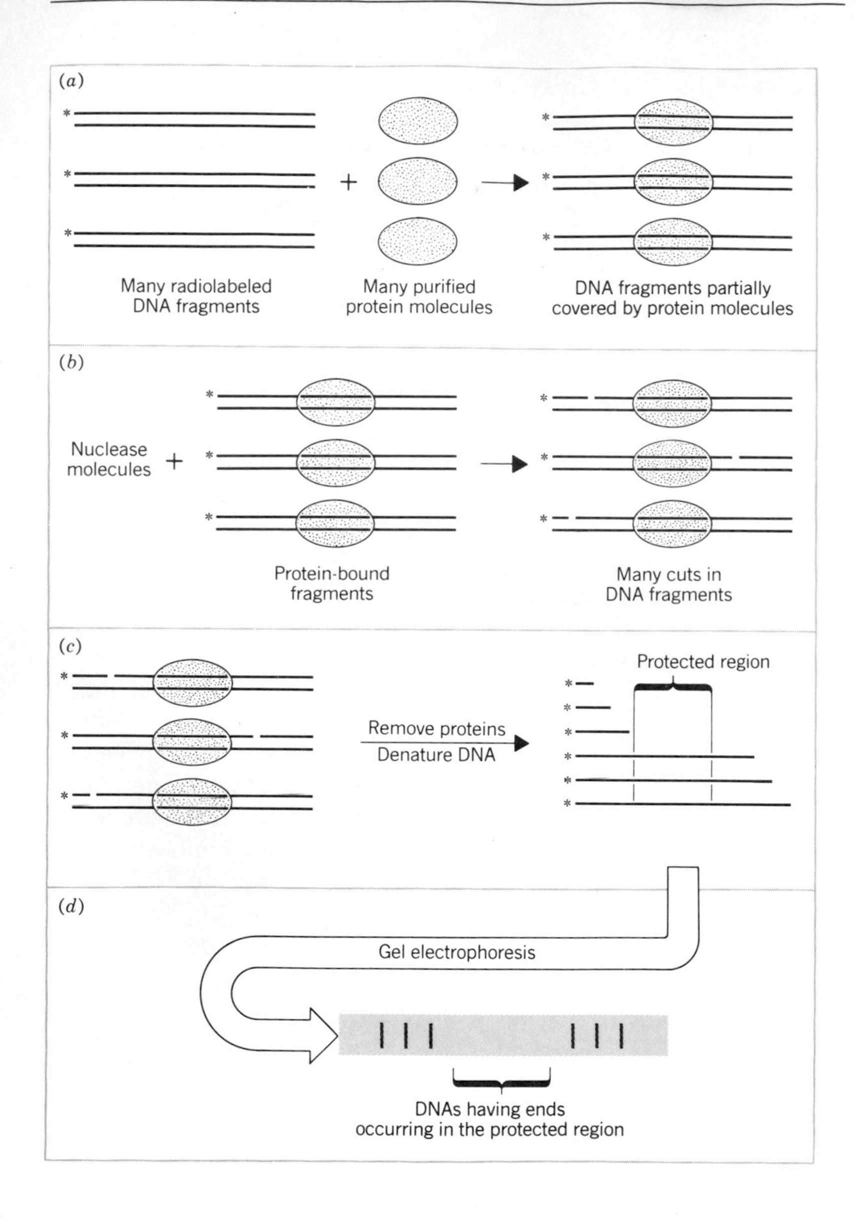
(a)
Many radiolabeled DNA fragments
+
Many purified protein molecules
DNA fragments partially covered by protein molecules
(b)
Nuclease molecules
+
Protein-bound fragments
Many cuts in DNA fragments
(c)
Remove proteins
Denature DNA
Protected region
(d)
Gel electrophoresis
DNAs having ends occurring in the protected region

To understand cell division, and most other cellular processes, we must know how the products of the relevant genes work and interact.

Gene cloning may have its greatest impact on biology by helping us understand the interactions among gene products, for cloning makes it much easier to apply both biochemical and genetic techniques to a single problem. These two approaches produce different, but complementary types of information. Biochemistry yields precise information about how molecules behave in a test tube, but whether they act in the same way inside living cells is a guess. Genetics provides information about how a molecule works in a living cell, but the information is imprecise.

Genetics utilizes mutations to perturb the normal function of a gene in a living cell. As the effects of the mutations are examined, ideas develop about how the genes and their products are interacting. For example, when a mutation in a particular gene causes DNA synthesis to stop, we know that the product of that gene is somehow involved in DNA synthesis. Gene cloning allows us to purify large amounts of the gene product so that its interactions with other molecules can be studied biochemically. Then we can begin to learn exactly how DNA synthesis works. Gene cloning also allows us to purify specific genes whose products we may have already studied biochemically. We can then create a mutation in the gene and put the mutated gene into a liv-

---

Figure 9-10 Protection of DNA by a Protein. (**a**) DNA fragments that have a radioactive label (asterisk) on one end of one strand are mixed with purified protein molecules. The proteins bind to the DNA fragments, covering a short region of DNA. (**b**) Nuclease molecules are added to the protein–DNA complexes, and the DNA strands are cut. Cutting can occur anywhere except where the protein is bound to the DNA. (**c**) The proteins are removed by detergent and protease treatments, and the double-stranded molecules are converted by mild heat or alkali treatment into single-stranded ones. The nuclease treatment in (b) causes the DNAs to have many different lengths; only radioactive DNAs are considered because only radioactive ones are measured below in (d). Note that not all DNA sizes are present because the protein blocks the cutting in certain regions during the nuclease treatment in (b) to produce a protected region of DNA. (**d**) The lengths of the radioactive DNAs are analyzed by gel electrophoresis; no bands are observed that have ends occurring in the protected region. In a sense, the protein has left its footprint on the DNA, and this type of analysis is often called footprinting.

ing cell. By examining the effects of the mutation and by combining that information with biochemical data, we can begin to understand how the gene functions. One of the more successful uses of this strategy has been in the study of transposition, a subject that is briefly outlined near the end of the next chapter.

Questions for Discussion

1. Gene cloning technologies, especially the polymerase chain reaction, can be used to identify people using skin fragments and other tissue samples. If you were a juror in a murder trial in which the primary evidence was based on polynucleotide chain reaction analysis, would you be able to vote guilty, remembering that there must be no reasonable doubt?
2. The restriction map of the *E. coli* chromosome (more than 4 million base pairs) is now nearly complete. To get this information, DNA fragments representing virtually the whole genome have been cloned into bacteriophage lambda, and a large collection of these recombinant phages is available. How could you use this information to set up a system to determine the map location of any *E. coli* DNA fragment that you happened to clone?
3. One practical application of gene cloning is detection of genetic abnormalities in fetuses. Develop a strategy to detect a disease that causes a change in the size of a specific restriction fragment using Southern hybridization. What materials (cloned genes, etc.) must be available to you? How could your system be used to detect carriers of the disease?
4. RNA molecules can be synthesized that have a nucleotide sequence complementary to a given messenger RNA. Such a complementary RNA molecule is called **antisense RNA**. How might antisense RNA be useful in fighting virus infections?

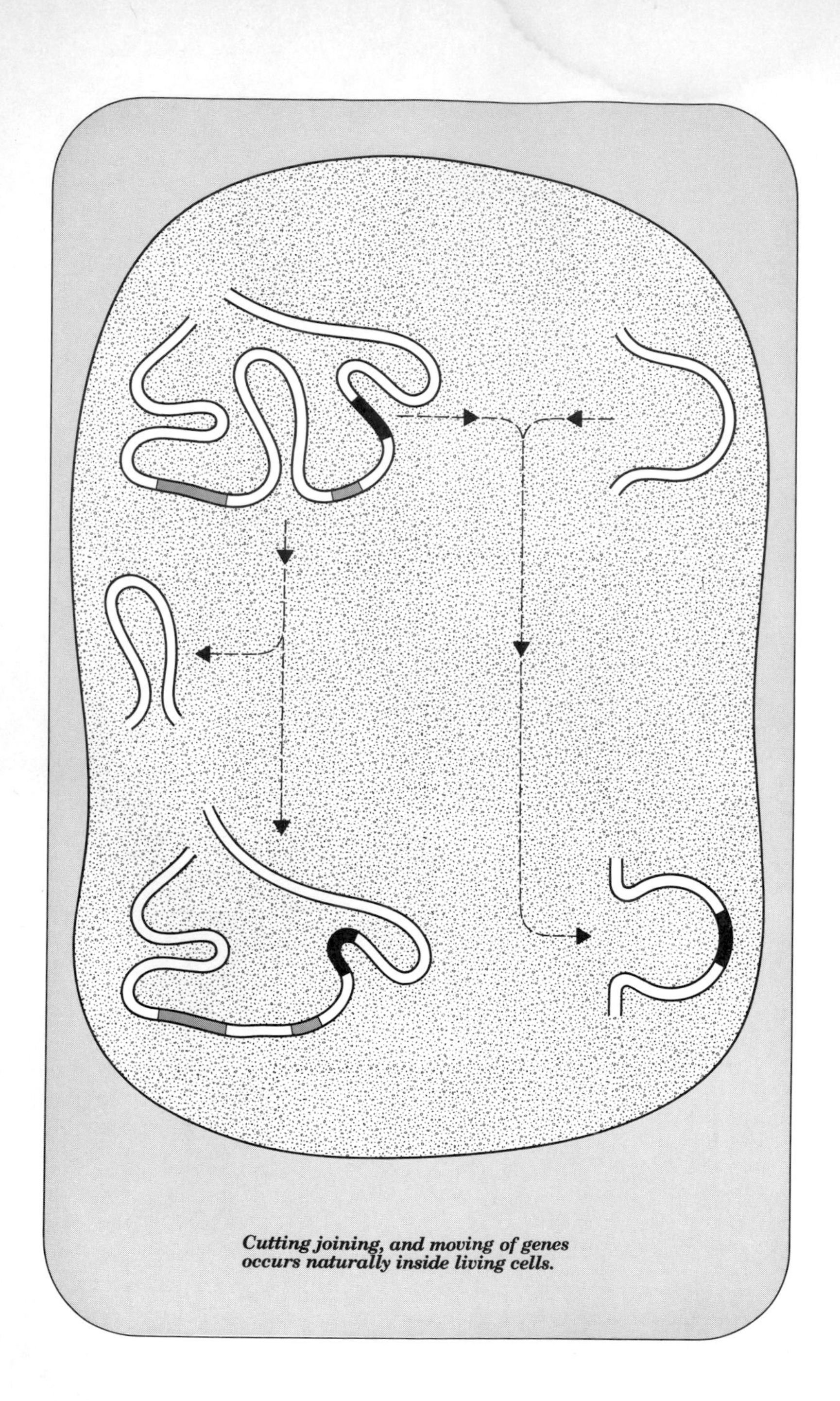

Cutting joining, and moving of genes occurs naturally inside living cells.

CHAPTER TEN

NEW PERSPECTIVES

A Sampling of Recent Insights into the Life Process

Overview

Gene cloning technologies have allowed us to purify large quantities of specific regions of DNA. The precise locations of many genes have been established by determining the nucleotide sequences of these regions. These studies have confirmed that the general biochemical principles of life are the same in humans and in bacteria. However, some remarkable differences exist. Unlike bacterial DNA, our DNA contains nonfunctional copies of genes. Moreover, our functional genes contain long stretches of nucleotides that code for nothing. They are excised from the transcript before it is translated into protein. The sequencing studies have also demonstrated that cutting, moving, and joining specific sections of DNA are normal processes that occur inside living cells.

INTRODUCTION

The first nine chapters outlined the principles of molecular genetics and gene cloning. Recent developments can now be described. This chapter offers six examples; many others are described in the papers listed as Additional Reading. The six stories chosen for this chapter illustrate how dynamic nucleic acids are. The first case, describing the globin gene family, serves to introduce the concept of **pseudogenes**.

Pseudogenes appear to be relics of once-active genes, and their existence raises questions about gene duplication and evolution. The second and third cases concern RNA splicing. It is now clear that RNA can cut and splice itself; therefore enzymes may not always be protein molecules. The fourth case, antibody formation, illustrates how gene rearrangements play a key role in the development of immunity. The fifth case focuses on transposition, the movement of small parasitic stretches of DNA from one region of DNA to another. The sixth describes an intimate relationship between a bacterium and a plant cell in which a portion of a bacterial plasmid is transferred to the plant genome, producing a tumor and indicating a way to engineer species of plants.

HEMOGLOBIN GENES AND PSEUDOGENES

The blood protein **hemoglobin** has been extensively studied for many years, and a number of statements can be made about it. First, hemoglobin is composed of four subunits, four separate protein chains called **globins**, that spontaneously associate to form the active protein. The four separate protein chains are of two types, alpha (α) and beta (β). The two types of protein, which differ slightly in length and in amino acid sequence, are paired in the hemoglobin molecule. Thus the predominant adult hemoglobin is generally called $\alpha_2 \beta_2$ (Figure 10-1). Second, several kinds of hemoglobin exist, and at different stages of life our genes instruct our blood cells to produce different globin proteins (see Figure 10-2). Each kind of hemoglobin is distinguished by having subunit chains of different types. For example, the blood of young human **embryos** contains two kinds of embryonic hemoglobin, $\alpha_2 \epsilon_2$ and $\zeta_2 \epsilon_2$. After 8 weeks of gestation the embryonic

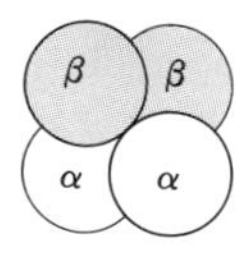

Figure 10-1 Structure of Hemoglobin. Hemoglobin is composed of two each of two types of protein subunit. In the predominant form of adult hemoglobin, these subunits are called alpha (α) and beta (β).

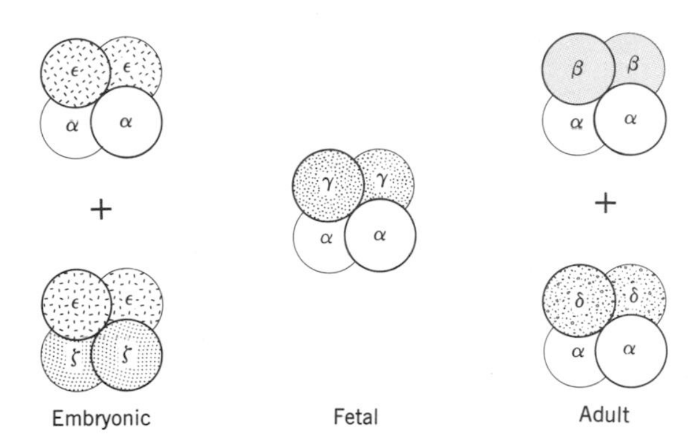

Figure 10-2 Hemoglobin Changes During Development. As humans develop from embryo to adult, the proteins that make up hemoglobin change (the names of the proteins are indicated by Greek letters).

forms are gradually replaced by the **fetal** form of hemoglobin, $\alpha_2 \gamma_2$. Fetal hemoglobin, which predominates until about 6 months after birth, is replaced by the adult forms, $\alpha_2 \delta_2$ and $\alpha_2 \beta_2$. A third statement is that separate genes code for each of the hemoglobin subunits, ξ, α, ϵ, γ, δ, and β. Thus we are faced with the question of how the various genes are switched on and off during our development to produce the correct kinds of hemoglobin for each stage.

Recombinant DNA technologies and DNA sequencing studies have not yet answered how gene switching occurs, but they do allow us to make four more statements about how the genes are organized (see Figure 10-3 for schematic drawing of globin gene arrangement). First, globin genes fall into two classes. The α class includes genes α and ξ, while the β class includes genes β, γ (subtypes G and A), ϵ, and δ. Second, the genes in one class are located in the same region of DNA; but members of the other class are located far away on another chromosome. Third, the proteins in one class, and thus the genes that code for them, have similar structures. For example, all the protein products from the embryonic and adult α-class genes are 141 amino acids long and vary only slightly in amino acid sequence. Fourth, the human genes in a class map in the same order in which they are expressed during development. For example, the β class genes map in order ϵ (embryonic), γ^G (fetal), γ^A (fetal), δ (adult), and β (adult) (Figure 10-3). Order is also preserved in the direction of synthesis of messenger RNA

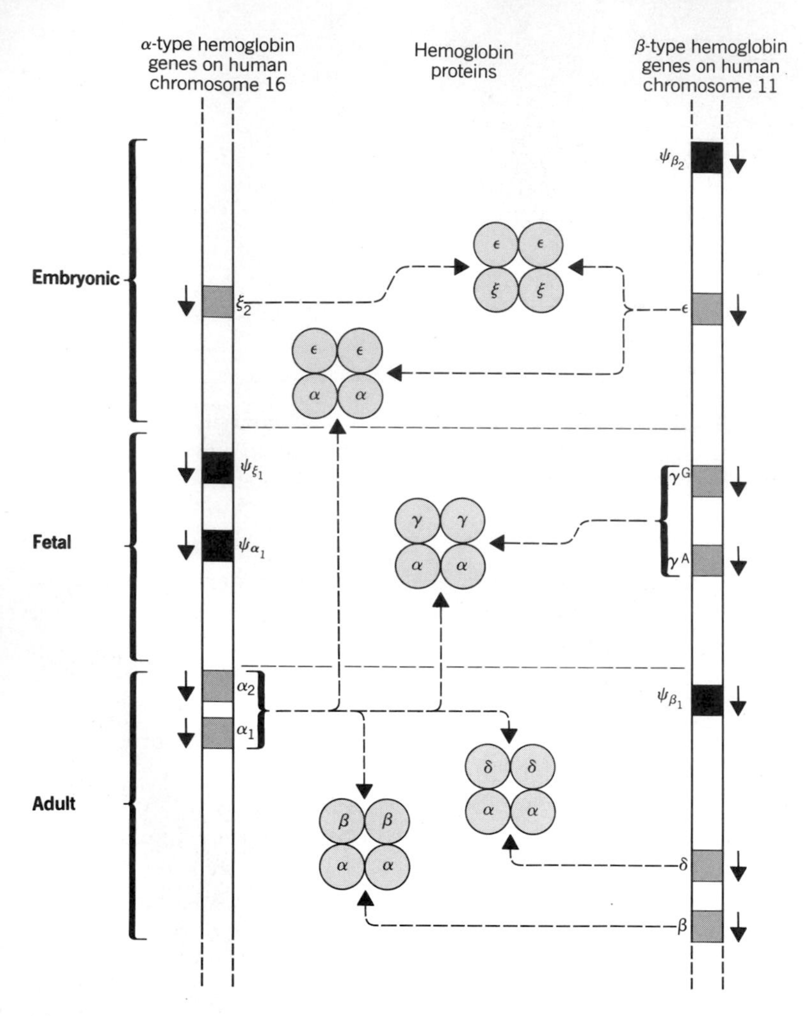

Figure 10-3 Human Globin Genes and Their Protein Products. Hemoglobin (depicted as four circles) is composed of two types of protein, and the composition varies with the stage of development through differential activation of globin genes. The embryonic forms (**a**) predominate until 8 weeks of gestation, the fetal form (**b**) to 6 months after birth, and the adult forms (**c**) from 6 months on. Heavy arrows indicate the direction of transcription, solid areas represent pseudogenes (nonfunctional genes; see text). The β-type genes are scattered over a stretch of 52,000 nucleotide pairs, whereas the α-type genes fall within a region containing 36,000 nucleotide pairs.

from the genes. For each gene, RNA synthesis starts at the end closest to the embryonic gene (arrows, Figure 10-3), and all the messenger RNAs are made from the same strand of DNA (the two DNA strands are complementary, not identical, and they would not code for the same proteins). Why this orderliness exists may become clearer when we understand how the developmental switches in globin gene expression occur.

Sequencing studies of the globin gene regions also uncovered several other short stretches of DNA that look remarkably like bona fide globin genes. Careful analysis of the nucleotide arrangements showed that these "genes" (solid regions, φ in Figure 10-3) are incapable of producing functional globin proteins—they are full of mutations and abnormalities. Premature stop codons, frameshift mutations, abnormal RNA polymerase binding sites (promoters), faulty initiation codons, and large internal deletions ensure that no functional globin proteins can come from these genelike regions called pseudogenes. The study of the hemoglobin genes and the surprising discovery of pseudogenes confronts us with a new set of questions. Did pseudogenes arise from the duplication of a preexisting gene? Did all the modern globin genes arise by duplication of a primitive ancestor? How does gene duplication occur?

The hemoglobin system is complicated, and, as in all complicated biological systems, there are many steps at which things can go wrong and produce serious diseases. One class of disease arises from nucleotide substitutions in the genes, causing amino acid changes in the globin proteins. Of the 300 or so mutations that have been identified, the most widely known is the mutation in the β-globin gene that causes sickle-cell anemia (see Figure 5-5). Another group of hemoglobin disorders, called the thalassemias, comes from a deficiency of one specific type of globin protein. Still other problems arise when the normal switch from one form of hemoglobin to another fails to occur.

EXONS, INTRONS, AND RNA SPLICING

Once genes had been purified, it was a straightforward process to determine the nucleotide sequences of the genes and of the DNA surrounding them. This led to the astounding discovery that in higher organisms the coding regions of many genes are interrupted by long

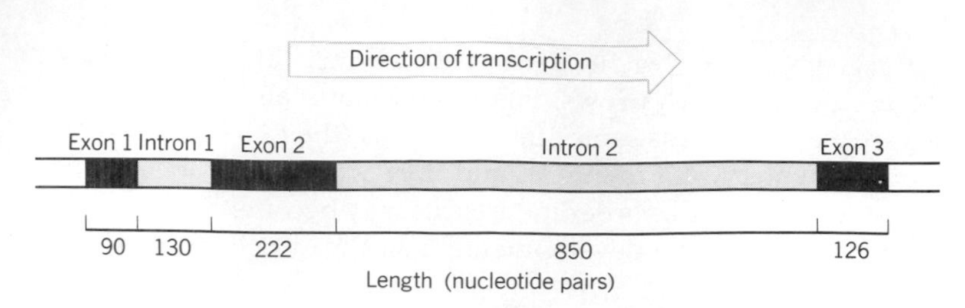

Figure 10-4 Intervening Sequences in the β-Globin Gene of Humans. The exons code for the amino acids in the protein.

stretches of nucleotides that do not code for amino acids found in the protein. The organization of the human β-globin gene is shown in Figure 10-4 as a gene containing three coding regions, called **exons**, interrupted by two noncoding regions, or **introns**. Cases of more than 50 introns scattered through a single gene have been reported. The RNA molecules transcribed from genes containing introns are longer than the messenger RNAs that subsequently produce the protein specified by the gene (Figure 10-5). Cells therefore have a mechanism that removes introns and yields mRNA composed only of coding sequences (exons). This process is called **RNA splicing**.

Although an intron-containing phage gene has been found, in general bacterial genes do not contain introns; thus splicing is not a part of bacterial messenger RNA processing. The reason for this difference between bacteria and higher cells is not known. Certainly RNA splicing provides many additional options for controlling gene expression. This is particularly apparent with animal viruses, where different patterns of splicing can allow a single region of DNA to produce several different proteins. Splicing is also important in generating antibodies of different types, as will be discussed later in this chapter.

RNA AS AN ENZYME

Earlier chapters stressed that proteins, as enzymes, control cellular chemistry by accelerating specific biochemical reactions. Nucleic acids, on the other hand, were portrayed as repositories for genetic information, repositories whose general function is to direct the synthesis of all cellular proteins. Thus it was a surprise to find that certain RNA molecules can act as enzymes.

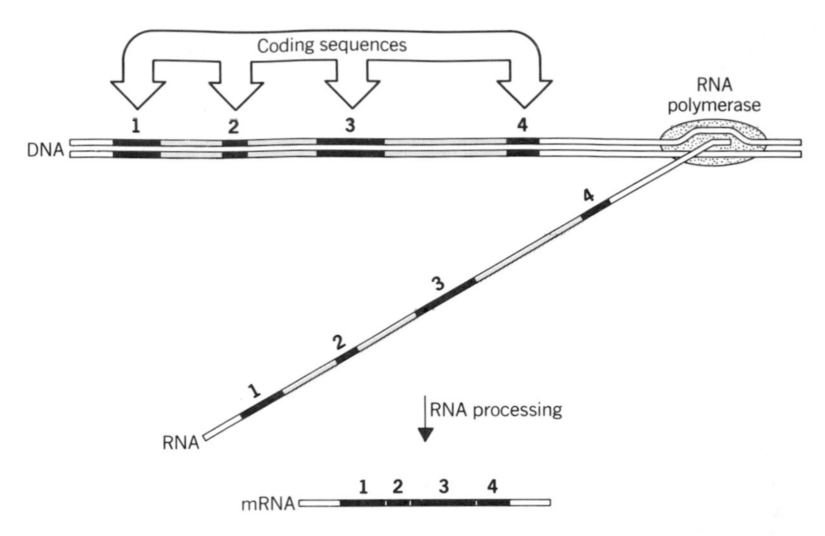

Figure 10-5 Arrangement of Sequences in a Gene from a Higher Organism. The regions of DNA coding for amino acids in the protein product are interspersed with noncoding regions, which are processed out (spliced) during formation of mature mRNA. Coding sequences **1** through **4** are part of a single gene. Note that in higher organisms ribosomes do not bind to messenger RNA before it is released from the DNA.

The first example of an RNA enzyme was found in ribosomal RNA of a protozoan, a small unicellular animal. Ribosomal RNA molecules are transcribed from DNA as much longer molecules that are later cut and processed. RNA splicing occurs during the processing, and excision of a particular intron was found to occur even after all proteins had been removed. The intron RNA is self-splicing. After excision, the intron retains its splicing capability and will act on other small RNA molecules. With certain oligonucleotides as substrates, the intron even makes oligonucleotides longer. Thus in a sense RNA can synthesize RNA as well as cut it. Such catalytic RNAs are called ribozymes.

The ability of RNA to catalyze chemical reactions has important biological implications. One is that RNA, which is a major component of ribosomes, may play an active role in the joining of amino acids to form proteins. Another concerns the origin of life. Since RNA molecules can act as enzymes as well as informational molecules, perhaps when life began, RNA functioned without DNA or proteins.

Ribozymes also represent a new way to manipulate and control the biology of living cells—they can be used to cut specific messenger RNA molecules. To be a suitable target for one type of ribozyme, a messenger RNA need only contain the nucleotide sequence GUC. The ribozyme consists of a catalytic region, which is 22 nucleotides long, and short flanking regions that can be any nucleotide sequence. If a ribozyme is made with its flanking regions complementary to the RNA target on either side of the target GUC, the ribozyme will bind to the target by complementary base pairing. This will bring the catalytic region into position to cut the target RNA (Figure 10-6). Since an entire ribozyme can be synthesized in the laboratory, the nucleotide se-

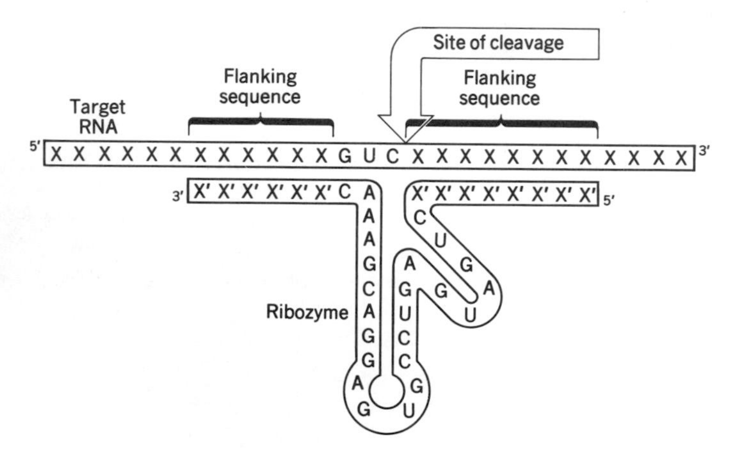

Figure 10-6 An RNA-Cleaving Ribozyme and Its Target. Partial complementarity allows the two RNA molecules shown to base-pair. Any RNA can be the target of the ribozyme shown, as long as the target contains the three-base sequence GUC (or a few other triplets). Cleavage occurs on the 3′ side of the C. This ribozyme is composed of a central catalytic region and two flanking regions. The flanking regions (nucleotides labeled X′) base-pair with complementary regions of the target (labeled X) and bring the catalytic region into proper position to cleave the target RNA.

A ribozyme of the type shown is constructed by first selecting a target region. This is done by scanning the RNA nucleotide sequence of the target gene for a GUC sequence. A particular gene may have several such sequences. Once the target has been chosen, a DNA molecule is chemically synthesized to serve as a template for synthesis of the ribozyme. The DNA usually has sites for restriction nucleases at its ends, so it can be easily inserted into a cloning vehicle for delivery into a cell. There RNA polymerase will properly synthesize the ribozyme.

quence in each region can be varied at will. Thus a ribozyme can be made that will attack any RNA that contains a GUC. The ribozyme can then be delivered to a cell by cloning the DNA template for the ribozyme into a plasmid or virus immediately downstream from a promoter. The intracellular synthesis of the ribozyme, and thus destruction of its target, can then be regulated by controlling transcription from the promoter (see Chapter 4 for a discussion of transcriptional control). Ribozymes are very specific, and if they are not directed at an RNA essential for cell growth, they cause little harm to the cell. Thus in principle they could be used to protect cells from virus infection or to rid cells of harmful gene products.

ANTIBODIES

Antibodies are proteins that recognize and bind to foreign substances, that is, substances not normally found in our bodies. As such, antibodies form part of our immune system, the elaborate network of molecules and cells that protects us from many types of disease. When an antibody attaches to a foreign substance, which is called an **antigen**, a number of processes that result in destruction or expulsion of the antigen are activated in the immune system. Millions of antigens can be recognized by antibodies. Since each different antigen is recognized by a different antibody, our bodies must be able to produce millions of different types of antibody.

Each antibody is composed of four protein chains, two identical **heavy chains** and two identical **light chains**. The chains are folded and connected to form a "T" as shown in Figure 10-7. Comparison of amino acid sequences from many different antibodies has revealed several interesting features. First, antibodies can be grouped into classes based on the amino acid sequence and properties of the heavy chains. Second, within a class there are sections of the protein chains that are identical from one antibody to the next. These sections are called **constant regions**, and they determine the behavior of the antibody in our bodies. For example, heavy chain antibodies with one type of constant region circulate in the blood, those with another type attach to the surface of the cell that produced them, and still others bind to specific cells that release histamines. The third point is that each light chain, as well as each heavy chain, has regions of amino acids that are unique to that antibody. These regions are called **variable re-**

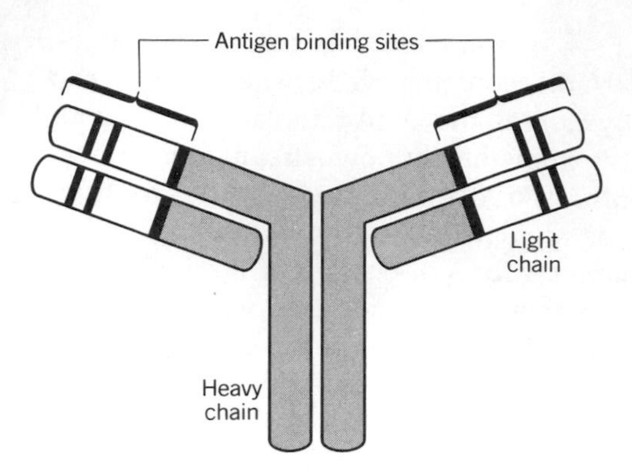

Figure 10-7 Schematic Diagram of an Antibody Molecule. Two heavy chains pair with each other and with two light chains to form the active antibody. The amino acid sequences are divided into constant regions (shaded), variable regions (open), and hypervariable regions (solid). Two antigen binding sites are present, one in the variable region of each arm.

gions; they are the parts of the antibody that bind to foreign substances such as viruses and bacteria. Since the shape and structure of a protein are dramatically affected by small changes in the sequence of amino acids, the slight differences in amino acids found in the variable regions result in millions of different antibodies, each able to recognize a particular antigen.

For several decades biologists were puzzled about how so many different antibodies could be produced. Could there be millions of genes, one for each protein chain of every antibody? Calculations suggested that we might not have enough DNA to code for all the antibodies and still have enough genes to run the chemistry of our cells. DNA sequencing studies now reveal that most of our cells do not have a complete set of antibody genes. Instead, they have bits and pieces that can be combined in a number of different ways, thus producing millions of distinct antibodies from a small amount of genetic information. The rearrangements occur inside special blood cells called **B lymphocytes**, which are responsible for making antibodies.

By comparing the nucleotide sequences in DNA from embryonic cells with those in DNA from antibody-producing cells, it has been possible to develop a general idea about how gene shuffling creates a variety of antibody chains. In the case of light chains (Figure 10-8), the

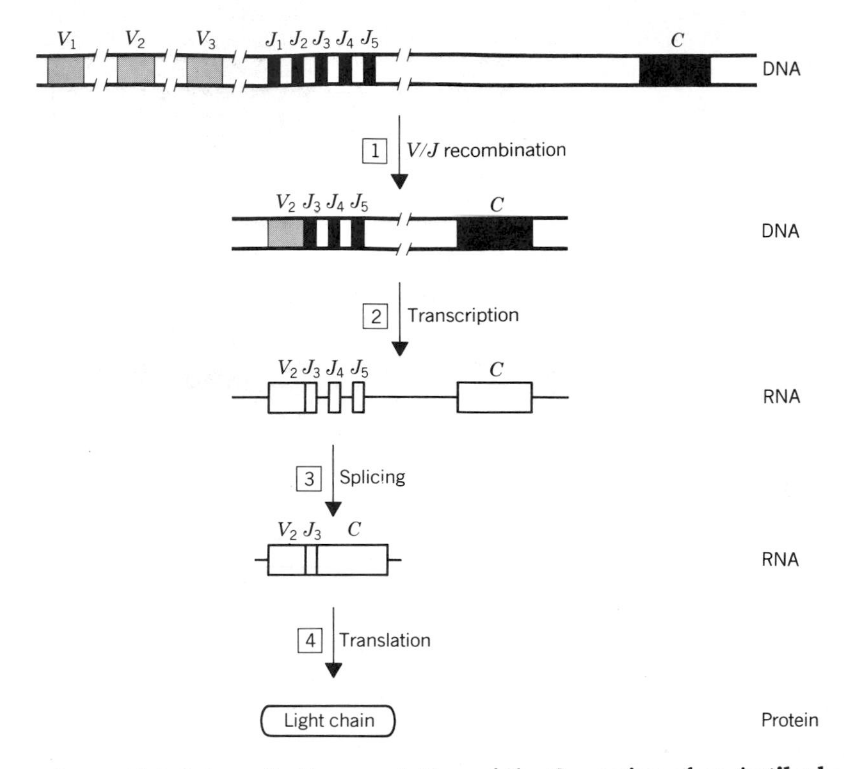

Figure 10-8 Schematic Representation of the Formation of an Antibody Light Chain. (1) One of the approximately 150 variable genes *V* recombines with one of the 5 joining genes *J*. In the example V_2 is moved and becomes adjacent to J_3. **(2)** RNA is synthesized from this DNA to produce a primary transcript. **(3)** Splicing occurs to remove all the RNA between J_3 and the constant gene C, producing mature messenger RNA. **(4)** This messenger RNA is translated into the antibody light chain. Discontinuities in the DNA indicate large distances between the genes.

embryo contains several hundred variable region genes (V) widely separated from five short, joining genes (J). DNA breakage and rejoining occur so that one of the V genes is placed next to one of the J genes. RNA polymerase transcribes this region and continues until it also transcribes a constant region gene (C). This long RNA molecule is then spliced to remove the sequence between the V/J region and the C region, producing mature messenger RNA. The messenger RNA is then translated into an antibody light chain. Since any one of perhaps 150 V genes can join to any of 5 J genes, roughly 750 combinations (150 × 5) can occur. Moreover, the joining sites are not precisely located; thus, the actual number of possible combinations is probably closer to 7500.

The same principles apply for heavy chain formation. However, more elements are involved in creating heavy chain diversity (Figure 10-9). In humans there are about 80 V (variable) genes, 50 D (diver-

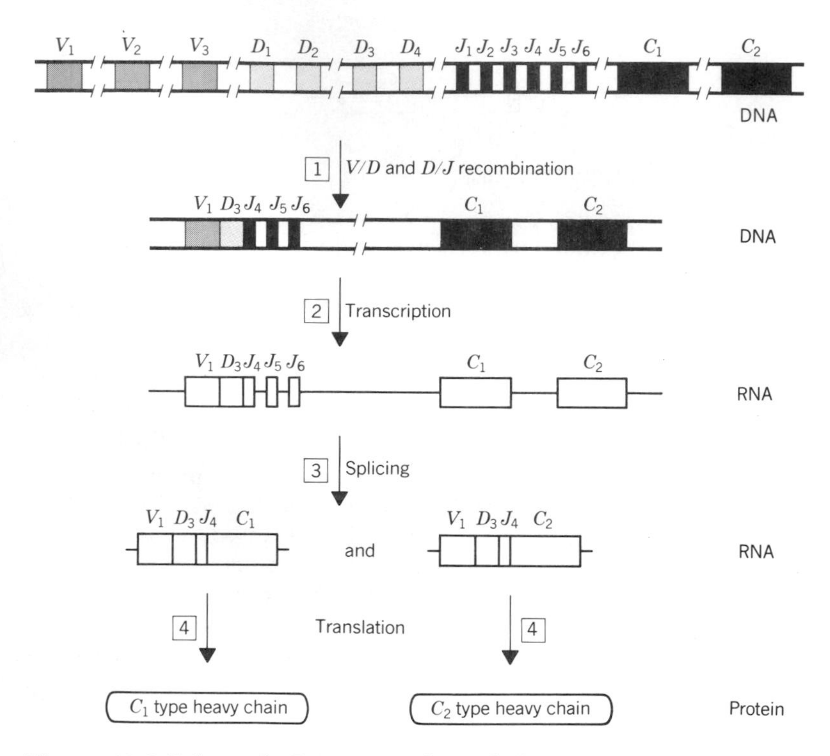

Figure 10-9 Schematic Representation of the Formation of Antibody Heavy Chains. **(1)** One of 80 *V* regions joins with one of about 50 *D* regions and one of 6 *J* regions to form a recombined DNA molecule in a cell called a B lymphocyte. **(2)** A primary transcript is made that contains two different **C** regions. **(3)** By differential splicing, two types of heavy chain messenger RNA can be made. **(4)** When the messenger RNAs are translated, they produce two types of heavy chain protein. Since the $V_1/D_3/J_4$ regions are the same for both, the two heavy chain proteins will have identical antigen binding sites. Discontinuities in the DNA indicate large distances between the genes.

sity) genes, and 6 J (joining) genes. Thus there are about 24,000 combinations (80 × 50 × 6) that can form. Flexibility in the V/D and the D/J junctions probably adds 100 more ways to combine the genes, so the total number of heavy chain combinations is about 2.4 million (24,000 × 100). The total number of antibody combinations is the product of the light chain and heavy chain combinations, or 18 billion (7500 × 2,400,000). Thus enormous diversity can be produced by about 300 embryonic DNA segments.

As mentioned earlier, there are several types of heavy chain, each with a different constant region. These constant regions determine how the antibody will behave in the body. Constant region genes are arranged downstream from the J region, and by selective splicing and additional recombination it is possible to put the identical V/D/J region onto five different constant regions. Consequently, our bodies can contain a number of antibody types that recognize the same foreign substance but perform different functions to combat it.

In summary, we appear to combat foreign substances in the following way. Our blood has millions of B lymphocytes circulating in it, and during development and maturation, each B lymphocyte undergoes a slightly different DNA arrangement. Thus each one produces antibody molecules that are slightly different from those produced by other B lymphocytes. At different times a cell can produce antibodies that differ in their heavy chain constant regions. Thus the antibody can behave differently at different times. The function of one of these classes is to reside on the surface of the B lymphocyte that produced it. There the antibody acts as a sentry, waiting to intercept a specific foreign substance, an antigen. When the antibody on the cell encounters an antigen to which it can bind, a complex is formed between the antigen and the antibody. The complex then triggers that particular lymphocyte to multiply and to produce additional antibody molecules (see Figure 10-10). All antibodies from a particular cell line have identical variable regions, so they all recognize the same antigen. Since the constant regions vary, we end up with antibodies that can function in different ways to rid our bodies of the antigen. Moreover, a complex antigen, such as a virus, has many different parts. Each can be recognized by a different B lymphocyte; consequently, the proliferating population of B lymphocytes will produce a population of antibodies capable of recognizing many different aspects of the virus. This increases our chances of fighting off infection.

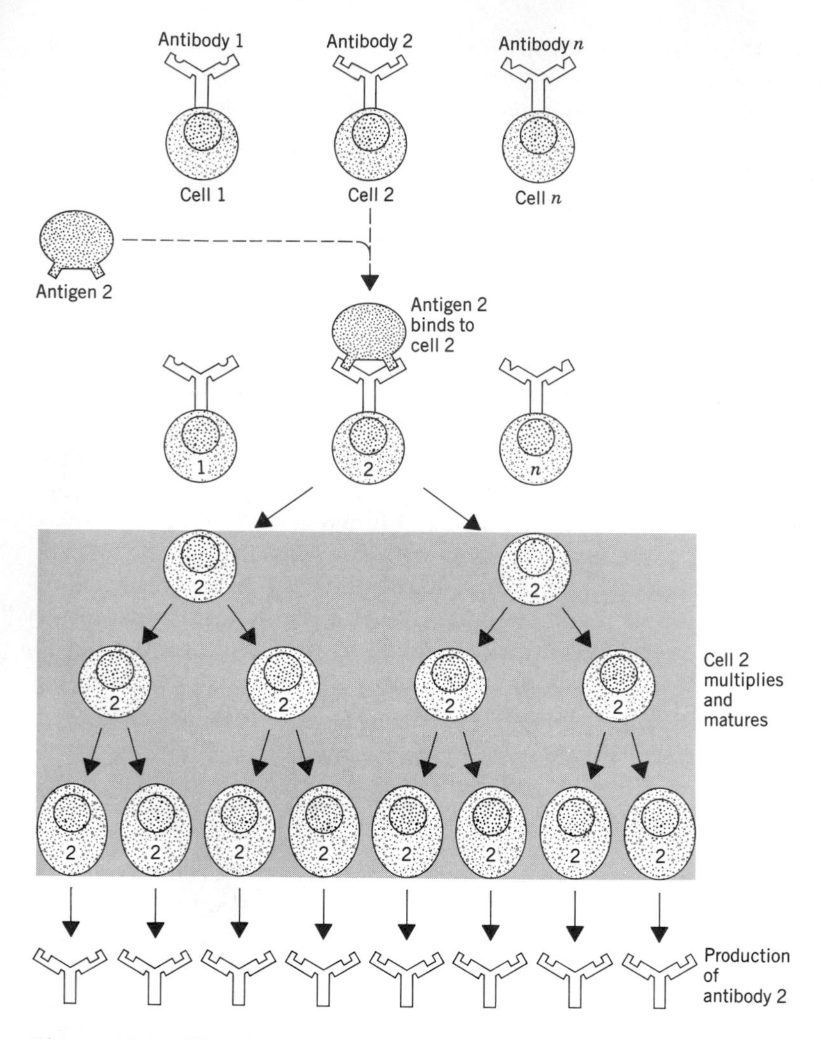

Figure 10-10 Clonal Selection for Antibody Production. Each antibody-producing B lymphocyte (cells 1, . . ., n) produces just one type of antibody, which is placed on the cell surface. Each B lymphocyte recognizes a different antigen. When an antigen comes in contact with a B lymphocyte in the proper way and is recognized by that cell, the cell is stimulated to divide and produce antibodies. This generates a large number of cells producing antibodies that recognize a specific antigen.

TRANSPOSITION

Gene cloning technologies allow us to insert small, discrete fragments of DNA into specific places in other DNA molecules. Nature also has this ability. Scattered among the genes of living cells are small, discrete sequences of nucleotides that can hop from one region of DNA to another or from one DNA molecule to another. These discrete nucleotide sequences are called **transposons**, and the process in which a transposon moves to another location in DNA is called **transposition**. In some types of transposition a duplication of the transposon occurs, as in Figure 10-11, while in other types the transposon excises and moves to a new location.

Transposition can have several consequences, depending on where the transposon inserts. For example, when a transposon inserts into the coding region of a gene, the information in the gene is interrupted (gene Y, Figure 10-11), and the gene may no longer produce a functional protein. Cases have also been observed in which transposition activates a gene by insertion near the gene.

Transposons have been found in a wide variety of organisms, and it is likely that all organisms contain them. Transposons from different

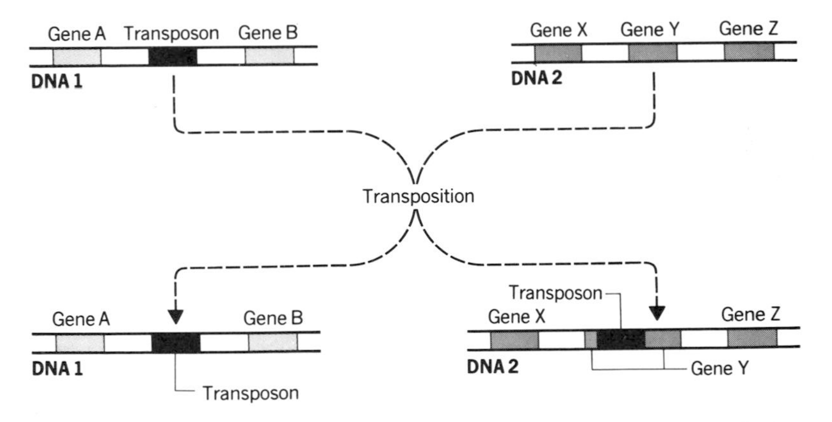

Figure 10-11 Transposition. A specific region of DNA 1 called a transposon (solid region) duplicates itself and inserts a copy into DNA 2. In this example the transposon inserts in the middle of gene Y, splitting the gene into two parts.

organisms appear to share three common features. First, transposons are always discrete sections of DNA; the junctions between a transposon and the DNA in which it is inserted are precisely defined. Second, transposons usually contain nucleotide sequences encoding one or more protein products required for movement of the transposon from one site to another. Thus transposons contain genes responsible for their own movement. Third, each end of a transposon contains nucleotide sequences that probably serve as recognition sites for factors involved in movement of the transposon. These sequences are often repeats of each other; in most cases the repeats are inverted. A general scheme of transposon structure is shown in Figure 10-12.

Although the molecular details of transposition vary from one type of transposon to another, a description of one called Tn3 is sufficient to provide an appreciation for the process. Tn3 is found in bacterial cells, and it contains three genes, *A*, *R*, and *bla* (Figure 10-13). In addition, Tn3 has a stretch of 38 base pairs at its left end that is repeated at its

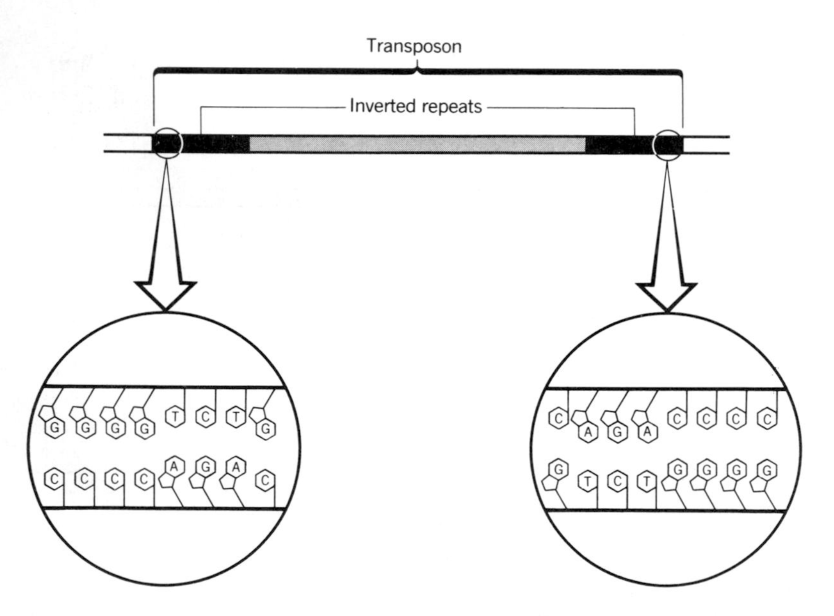

Figure 10-12 General Structure of a Transposon. Transposons contain repeated nucleotide sequences at each end. The repeated sequences are generally in an inverted orientation. Genes involved in transposon movement lie between or within the repeats.

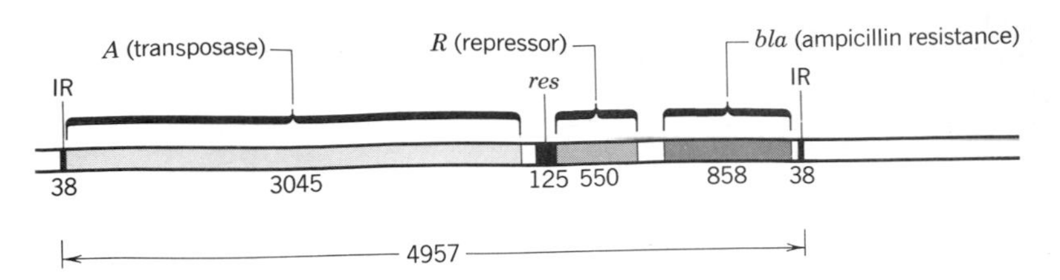

Figure 10-13 Arrangement of Genes in Tn3. Tn3 contains three genes, *A*, *R*, and *bla*, located between the 38-base-pair inverted repeats (*IR*). The repressor protein binds to Tn3 DNA at region *res*. Numbers indicate nucleotide pairs in the specified regions.

right end in an inverted orientation. The *bla* gene encodes a protein that destroys ampicillin (penicillin); thus any cell containing Tn3 is resistant to ampicillin. This feature helps biologists determine whether a cell contains Tn3. The *A* gene encodes a protein called **transposase**, a protein responsible for Tn3 movement. If small regions of the *A* gene are experimentally removed, Tn3 can no longer move. The *R* gene codes for a repressor that binds to the *A* gene and prevents it from making transposase. Thus the repressor keeps transposition from occurring very often.

It is likely that transposition of Tn3 occurs as a two-step process. In the first step (Figure 10-14) the DNA molecule containing Tn3 (the donor DNA) somehow binds to a DNA molecule lacking Tn3 (the recipient DNA). This process is mediated by the transposase protein. Then the Tn3 sequence is duplicated to produce a structure called a **cointegrate**. The cointegrate contains two copies of Tn3, one at each junction between the donor and recipient DNAs. After the cointegrate has formed, transposase dissociates from the DNA.

In the second step the cointegrate is split into two circles, and both donor and recipient DNA molecules end up with a copy of Tn3. The product of the *R* gene, the repressor, plays a key role in this process (Figure 10-15). It binds to the cointegrate DNA and aligns the two copies of Tn3. Breaks occur in the two copies of Tn3 at the *res* sites. One double-stranded DNA crosses over the other, and the broken ends are joined so that a part of one Tn3 is linked to the remainder of the other. This process of DNA strand exchange is called **recombination**; genetic information in the two copies of Tn3 has been ex-

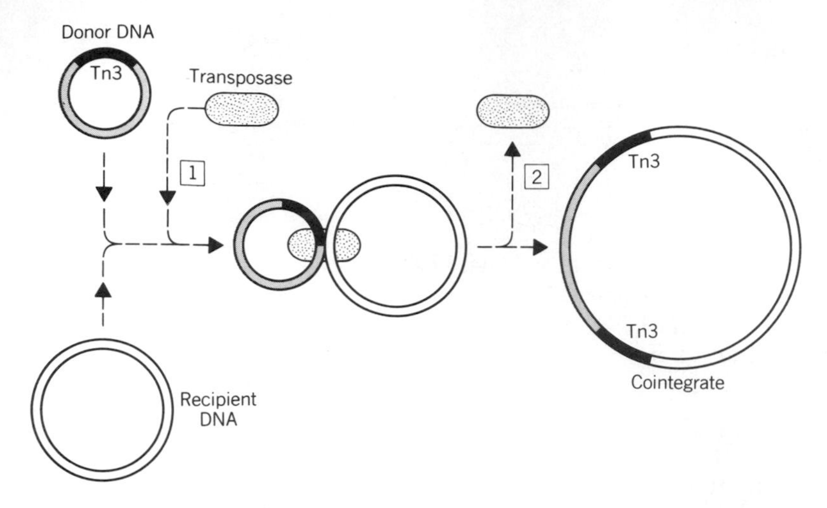

Figure 10-14 Scheme for Formation of Cointegrates by Tn3. **(1)** Transposase mediates the joining of a DNA molecule containing Tn3 (donor DNA) with a DNA lacking Tn3 (recipient DNA). **(2)** Tn3 is replicated (duplicated by DNA polymerase), and in the process the donor and recipient DNAs are joined to form a larger circle (cointegrate). The transposase leaves the DNA and is presumably free to initiate a new round of transposition.

changed. In this case recombination also creates two DNA circles, each with a copy of Tn3. Thus the repressor plays two roles in transposition of Tn3: first, it keeps the frequency of transposition low by repressing the *A* gene, and second, it is required to align, break, and rejoin cointegrate DNA. Another interesting feature is that the repressor also represses the *R* gene—it represses its own production. Thus transposition is a very tightly regulated process.

Transposons are among the most exciting tools for genetic research, and they offer a way to insert genes into animals. The following procedure has actually been carried out. A transposon from a fruit fly was cloned into a bacterial plasmid. Transposon-containing plasmid DNA was then isolated from the bacteria, and an eye color gene was inserted into the transposon. The engineered transposon, still on the plasmid, was returned to the bacterium, permitting investigators to obtain large quantities of the engineered transposon. Purified plasmids containing

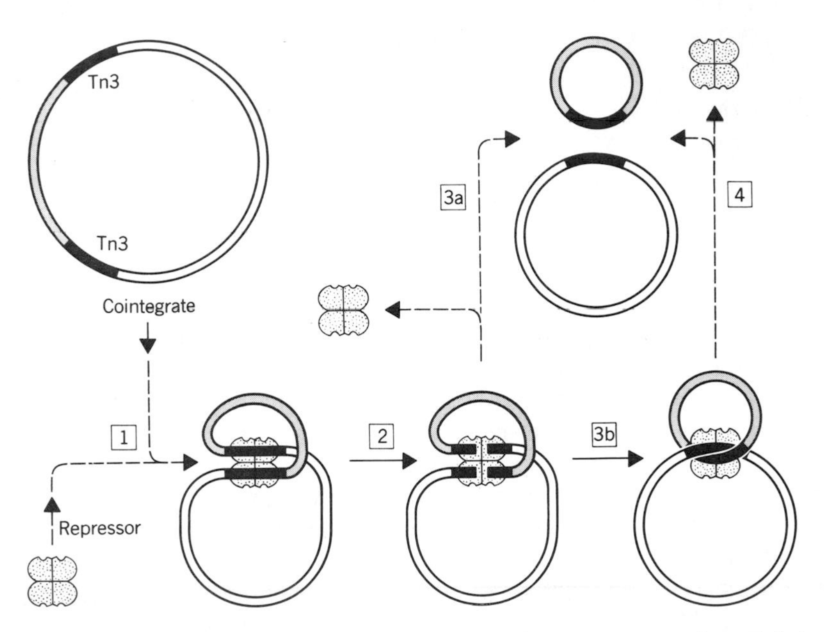

Figure 10-15 Recombination Between Two Tn3 Transposons in a Cointegrate. **(1)** The repressor protein binds to the cointegrate, and probably uses the natural twists in the DNA to align the two copies of Tn3 (solid). **(2)** A break occurs within the *res* site (see Figure 10-13) of each Tn3. **(3)** One DNA crosses over the other, and the breaks are sealed. This produces two rings, one of which is the donor (shaded) and the other is the recipient (open). Both contain a copy of Tn3. Recombination has two possible outcomes. **(3a)** Two separate rings are produced directly or **(3b)** two interlocked rings arise. In both cases the repressor dissociates from the rings **(3a, 4)**. **(4)** An enzyme called a **topoisomerase** will separate the interlocked rings.

the transposon were then injected into fruit fly embryos. There the transposon, along with the new eye color gene it carried, transposed into a fruit fly chromosome. When the fly developed into an adult, it had the eye color dictated by the gene that had been inserted into the transposon.

Transposons may also be important for their ability to inactivate genes. Some genetic diseases may be caused by transposons inactivating crucial genes, and if we had a way to remove the transposon, we might be able to reactivate the gene and thus cure the disease.

CROWN GALL TUMORS AND THE TI PLASMID

Plants can develop tumors, and an example arising from the crown gall disease is shown in Figure 10-16. Crown gall tumors are caused by a large bacterial plasmid called Ti whose normal host is a bacterium called *Agrobacterium tumefaciens.* The tumors result from an intimate relationship among the plant, the bacterium, and the plasmid. For tumor formation to occur, the living bacterium must gain access to the plant through a wound; then part of the Ti plasmid is transferred into the plant cells. The plasmid contains three genes encoding proteins required for the production of plant growth hormones; when these hormones are produced, the plant cells begin multiplying uncontrollably. The plasmid also contains genes (*nos*) that encode proteins required for the production of compounds called opines. Other genes (*noc*) on the plasmid specifically break down the opines into compounds that can be used by the infecting bacterium as a nutrient source. Thus the bacterium benefits from the uncontrolled cell growth and production of opines. The plant cells become genetically altered by the uptake of the plasmid DNA, and indeed tumor cells can be cultured indefinitely in the absence of the infecting bacterium or added growth hormones.

The Ti plasmid–*Agrobacterium tumefaciens* transformation system has recently been used to genetically manipulate plants. Up to 50,000 base pairs of foreign DNA can be inserted into the plasmid and subsequently incorporated into the plant DNA. This method has already been used to make both soybeans and tomatoes resistant to a herbicide.

PERSPECTIVE

DNA must be a stable structure, since this substance is the vehicle through which the characteristics of a species are accurately passed from one generation to the next. But DNA is not static, either in information content or in three-dimensional shape. As pointed out above, gene duplications probably occur, and over time mutations arise that lead to nucleotide sequence differences between duplicated genes. These differences may allow the protein products to perform different functions. Additionally, transposons hop around in the DNA, constantly creating the potential for changing gene structure. Even within

Figure 10-16 Crown Gall Tumor. A turnip root was inoculated a single time with a virulent strain of the bacterium *Agrobacterium tumefaciens.* Tumor cells initially developed at the site of inoculation and then spread. The bacteria can be removed from the tumor cells, and the cells can be cultured on agar. When inoculated into a healthy plant, the cultured tumor cells will generate a new tumor. (Photo courtesy of Dr. C. I. Kado, University of California, Davis.)

the lifetime of an individual, gene rearrangements can be quite common in some cell types.

Viruses and plasmids are responsible for another aspect of DNA dynamics, the movement of DNA from one cell to another. With viruses, gene transfer occurs when host DNA is mistakenly packaged inside virus particles and then carried to a new host upon infection by the virus. In bacteria some plasmids integrate (insert) into the host DNA. The plasmids then either mobilize the host DNA for direct transfer (see Appendix I) or improperly excise, with the result that a piece of host DNA remains a part of the plasmid. Release of the plasmid into the environment would allow it to infect a new host. The transfer of the Ti plasmid involves not only plasmid genes but also an intimate relationship between the bacterial and plant hosts. When **retroviruses** are involved, as is often the case with gene transfer in animals, the transfer includes a step in which the genes are in the form of RNA rather than DNA. There are even transposons whose biology is very similar to a degenerate retrovirus, one that no longer escapes from the cell. Because of their potential in human genetic engineering and their roles in AIDS and human cancer, retroviruses currently command a great deal of public attention.

Questions for Discussion

1. Why are people unlikely to show an immunological response to poison oak or poison ivy the first time they touch these plants?
2. If you were to compare nucleotide sequences between monkey and man, would you expect to find more differences in genes or pseudogenes?
3. Most human genes are not composed of a continuous stretch of DNA; instead they contain alternating stretches of coding DNA (exons) and noncoding regions (introns). Both the exons and the introns are transcribed into RNA, but the introns are removed via splicing. Sometimes the amount of DNA devoted to introns far exceeds that in exons. What purposes might the introns have that would justify the biochemical cost of synthesizing them?

4. RNA molecules called ribozymes can be synthesized that will specifically cleave other RNA molecules. One type of ribozyme cleaves next to the trinucleotide GUC. How might ribozymes be useful in fighting virus infections? What information would you need to know to design an effective ribozyme?
5. For a ribozyme to be effective against a virus, it must be present inside the cell. How might you generate a high concentration of a specific ribozyme inside a cell?

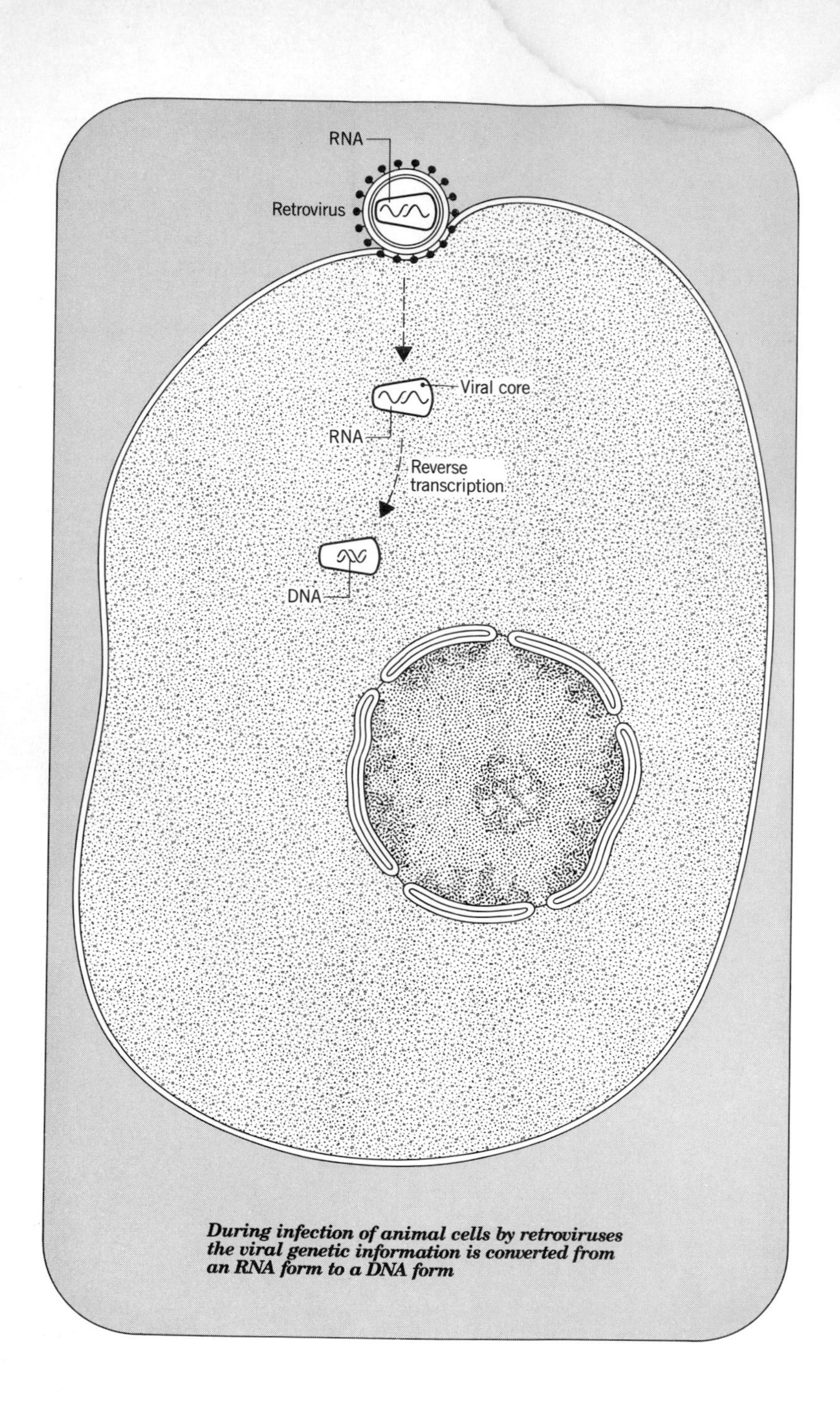

During infection of animal cells by retroviruses the viral genetic information is converted from an RNA form to a DNA form

CHAPTER ELEVEN

RETROVIRUSES

AIDS and Cancer Genes

Overview

Retroviruses comprise a group of related viruses in which the viral genetic information is converted from an RNA form to a DNA form during infection of animal cells. In its DNA form the viral information becomes inserted into the host chromosomes. From a chromosomal position, the viral DNA directs the synthesis of viral RNA and proteins, which then spontaneously assemble to form progeny virus. Viruses of this type have been detected in many animal species, and some have been popularly called RNA tumor viruses because of their association with tumors and leukemias. One retrovirus causes AIDS, killing cells normally involved in the human immune response. In addition to their role in disease, retroviruses serve as cloning vehicles and provide a source for reverse transcriptase, an enzyme valuable in genetic engineering.

INTRODUCTION

During the life cycle of retroviruses there is a stage in which genetic information flows from RNA to DNA, a direction opposite to that found in most genetic systems. This reverse, or retrograde flow of information is the basis for the name retrovirus. Once in the chemical form of DNA, the viral genetic information is inserted into the DNA of the infected cell. There it can remain for long periods of time as a molecular parasite. In principle, this insertion or integration process is similar to the formation of bacterial lysogens by bacteriophage lambda (see Chapter 6 and Figure 6-7).

Retroviruses are important for a number of reasons. The most important is that they cause disease in humans. One example is a deadly form of leukemia caused by a virus called HTLV-I, and another is the acquired immune deficiency syndrome (AIDS), caused by the Human Immunodeficiency Virus (HIV-1). AIDS is a lethal disease in which the immune system is debilitated to such a degree that it cannot protect the victim from microorganisms that are normally innocuous. Other retroviruses are called tumor viruses because they carry genes (oncogenes) that cause malignant cell growth.

Retroviruses are also used for gene cloning and genetic engineering. They are the source of reverse transcriptase, an enzymatic tool that makes it possible to synthesize DNA from RNA molecules. Moreover, retroviruses that have been altered to remove their pathogenic components are beginning to serve as cloning vectors for placing foreign genes into animal cells. Our understanding of these viruses and their interactions with their animal hosts is far from complete, and many paradoxes remain to be solved. But firm statements can be made about the molecular anatomy of the viruses and certain aspects of their life cycle. Some of these details are presented below to promote an understanding of the strategies for manipulating these viruses.

RETROVIRUS STRUCTURE

Retrovirus particles are spherical structures (Figure 11-1) containing two RNA molecules, each about 10,000 nucleotides long, and at least nine different types of protein (Figure 11-2). The RNA is packaged by viral proteins into a structure called a core. The proteins of the core, which are encoded by the viral gene called *gag*, are among the major constituents of the virus. The viral core also contains the reverse transcriptase needed to synthesize DNA from RNA, a site-specific protease, and a protein called **integrase**. Integrase participates in inserting the viral DNA into the host DNA. Surrounding the core is an **envelope** composed of a double layer of lipid (fatlike molecules) derived from the surface membrane of the host cell in which the virus was formed. Within the envelope are embedded proteins of several types which are encoded by the viral gene called *env*. Some of the viral proteins in the envelope specifically interact with cellular proteins called receptors.

These receptors are located on the surface of animal cells, and the interaction of the viral envelope proteins with cellular receptors enables the virus to infect specific animal cells.

Retroviruses have a common genetic organization (Figure 11-3*a*). At the ends of the viral RNA are blocks of repeated nucleotides called R. Internal to R at the 5′ end of the RNA is a block called U5; at the 3′ end is another block called U3. U5 and U3 are duplicated when the RNA is converted into the DNA form of the viral genome (see below), and the resulting DNA copy has 5′ U3-R-U5 3′ regions at each end. These sequences, called long terminal repeats (**LTRs**), are important because they contain the viral promoters from which viral RNA is synthesized. The outer ends of the LTRs contain nucleotide sequences crucial for insertion of the viral DNA into the host chromosome. The *gag* gene is nearest the 5′ end of the RNA, and it encodes a long **polyprotein** from which several proteins are subsequently cleaved. These proteins are part of the viral core, and they probably serve to wrap and protect the viral RNA. On the 3′ side of *gag*, and sometimes overlapping it slightly, is the gene called *pol*. *Pol* also encodes a long polyprotein, which is actually synthesized as a very long *gag-pol* polyprotein. In the case of HIV-1, the *pol* gene product is subsequently cut into three proteins: a specific protease responsible for cutting the polyproteins, reverse transcriptase, and integrase. The third gene, *env*, encodes a polyprotein which, when cleaved, produces two viral proteins that eventually become a part of the virus envelope. In the case of HIV-1 these two proteins are called gp120 and gp41 (gp stands for glycoprotein, i.e., sugar-containing protein; the numbers refer to the sizes of the proteins).

The genetic organization described above is common to all retroviruses, but variations on the theme occur. Viruses of the type that cause AIDS have several additional elements that control viral gene expression, and viruses that cause tumors tend to have parts of their genome replaced with animal genes. These special cases are described in more detail in later sections.

RETROVIRUS LIFE CYCLE

Our ability to grow animal cells in culture and infect them with viruses has made it possible to study the life cycles of many animal viruses including those that cause polio, influenza, and AIDS. In one method of

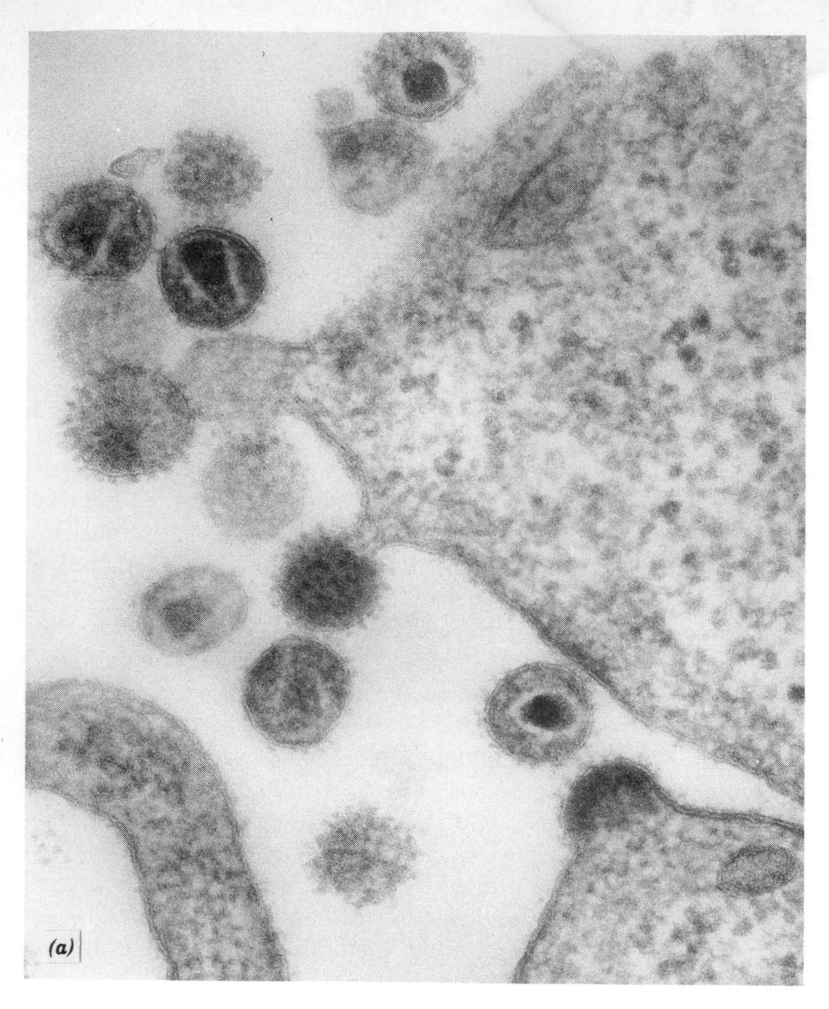

Figure 11-1 Retrovirus Structure. (**a**) Electron micrograph of HIV-1 budding from a human cell (magnification 120,000 times). Infected cells were cut into thin sections prior to microscopic examination. The particles reveal different features of viral structure because they were not all cut at the same place. [Photomicrograph courtesy of Dr. H. Gelderblom, Robert Koch Institute, Berlin, reprinted from *Arch. Virol.* **100**:255–266(1988) with permission.] (**b**) Diagrammatic representation of a retrovirus in cross section. The outer shell or envelope is composed of a double layer of lipid (solid) and protein. In HIV the major envelope glycoprotein has knobs that are called gp120, and this protein is anchored to the envelope by a transmembrane protein called gp41. The Env proteins exist in the membrane as **tetramers**. Inside the envelope is a core consisting of several types of protein from the *gag* gene that package and protect two RNA molecules. The position of the two RNA molecules and their orientation relative to each other are not known. Also present in the core are the viral reverse transcriptase and integrase. Adapted from W. Hazeltine and F. Wong-Staal, *Scientific American*, October 1988.

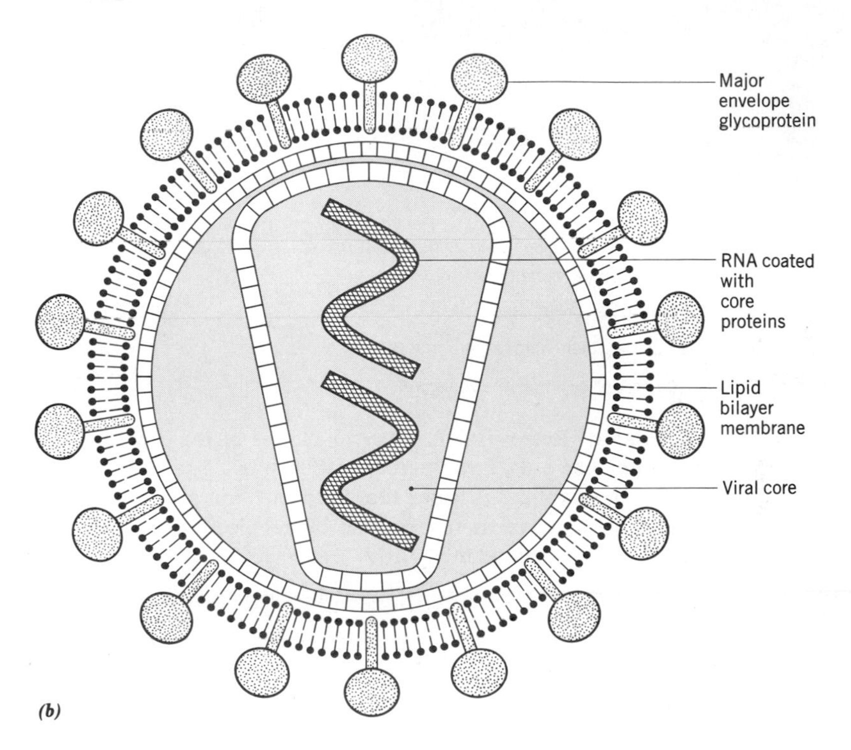

(b)

culturing cells a piece of tissue is first surgically removed from an animal. The tissue piece is then placed in a plastic petri dish containing a nutrient medium and treated with enzymes to detach the cells from each other. Some cell types grow and divide in suspension while others stick to the bottom of the dish. In the latter case, when the cells come into contact with each other, they stop growing and dividing. They tend to form a **monolayer**, a layer one cell thick on the bottom of the dish. In either case, dilution and transfer of the cells to another petri dish leads to renewed growth and division. In this way some types of animal cell can be perpetuated for many generations. Addition of virus particles to the cell culture results in infection and eventually in release of newly made virus into the culture medium. Some viruses cause infected cells to die. HIV-1, which infects cells growing in suspension, falls into this category. Cells infected with HIV-1 may also fuse with other infected cells to form giant cells containing many nuclei.

The first stage of retroviral infection is the entry of the virus into an animal cell (Figure 11-4*a*). This appears to occur by a fusion of the viral envelope with the cell membrane, leading to the movement of

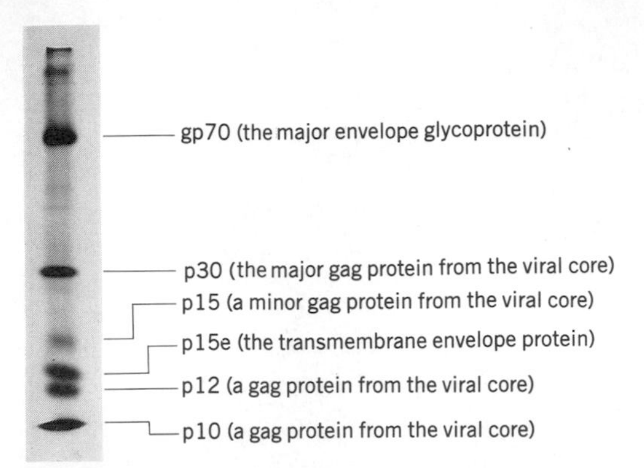

Figure 11-2 Display of Retroviral Proteins. Particles of the FrMCF virus were isolated from mouse cells, and the preparation was treated with detergent to disrupt the chemical interactions that hold the viral components together. The viral proteins were then separated by gel electrophoresis using methods similar to those described in Figure 7-4. An electric current forces the proteins to move through a gel of acrylamide. Under the conditions used, smaller proteins move faster than larger ones. In the figure, the direction of migration is from top to bottom. This particular preparation was radioactively labeled, and after electrophoresis the gel was used to expose X-ray film. Dark spots appear on the film where the proteins are located. The proteins have been given names based on their size; thus analogous proteins from different viruses often have different names. The letter "g" indicates that sugar groups are attached to the protein. Proteins such as reverse transcriptase, integrase, and the viral protease are present at such low concentrations that they do not show up with this type of analysis. (Photo courtesy of William Honnen and Abraham Pinter, Public Health Research Institute.)

the viral core into the cytoplasm of the cell (Figure 11-4*b*). Specific proteins on the cell surface serve as receptors for the virus and allow the penetration process to occur. Consequently, only certain types of cell are infected, those containing the receptor protein. For example, HIV-1, the AIDS virus, preferentially infects specific cells of the human immune system by interacting with a particular receptor protein, called CD4, which is present on the surfaces of these cells.

Soon after entry into the cell, the viral reverse transcriptase (a DNA polymerase) makes a double-stranded DNA copy of the information in the viral RNA (Figure 11-4*c*). The viral DNA then inserts into the host chromosome by a process called **integration** (Figure 11-4*d*). Both

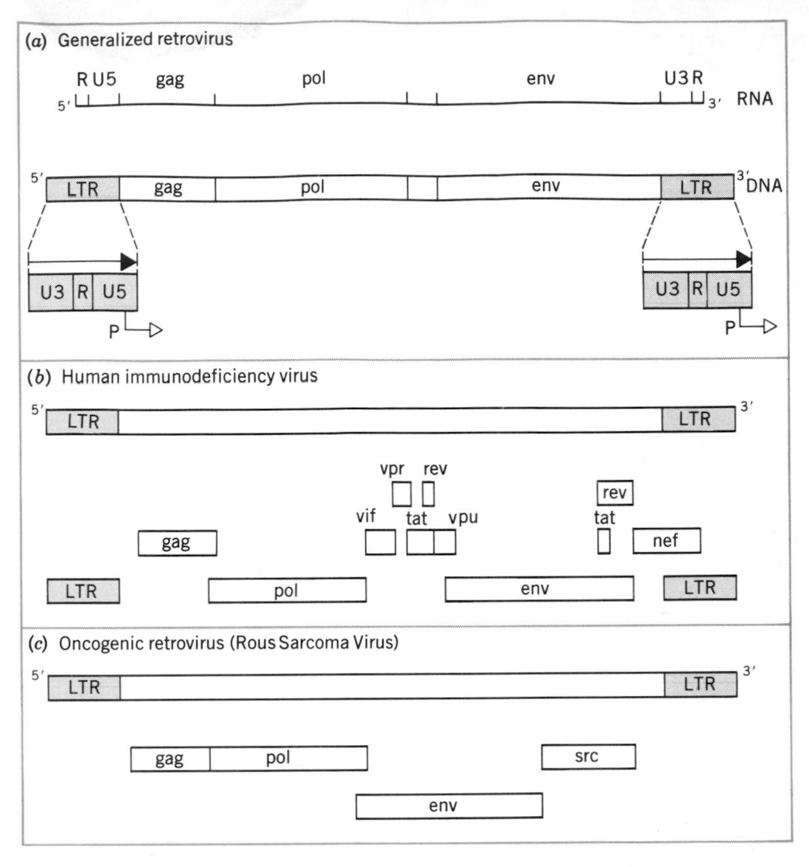

Figure 11-3 Genetic Organization of Retroviruses. (a) Although each retroviral species has a distinct nucleotide sequence, all retroviruses share a common genetic structure. The region called R is a repeat found at both ends of the RNA. Regions called U5 and U3 occur only at 5′ and 3′ ends, respectively, of the viral RNA, but they are both present at each end of the DNA. The U3-R-U5 arrangement creates two long, terminally repeated sequences (LTRs), and the DNA is longer than the RNA by the number of nucleotides in U5 and U3. The positions labeled **P** are strong promoters, and the direction of transcription is indicated by arrows. The three genes *gag, pol,* and *env* are common to all retroviruses. The 5′ and 3′ notations on the DNA are for the coding strand. **(b)** The AIDS virus (HIV-1) has the standard retrovirus organization plus additional genes called *vif, vpr, vpu, tat, rev,* and *nef*. Many of these genes overlap as indicated. The *tat* and *rev* genes are split, and their expression requires RNA splicing. **(c)** The oncogenic Rous Sarcoma Virus contains a gene called *src* on the 3′ side of *env*. The product of *src* is thought to perturb the control of cell division, leading to formation of a type of tumor called a sarcoma.

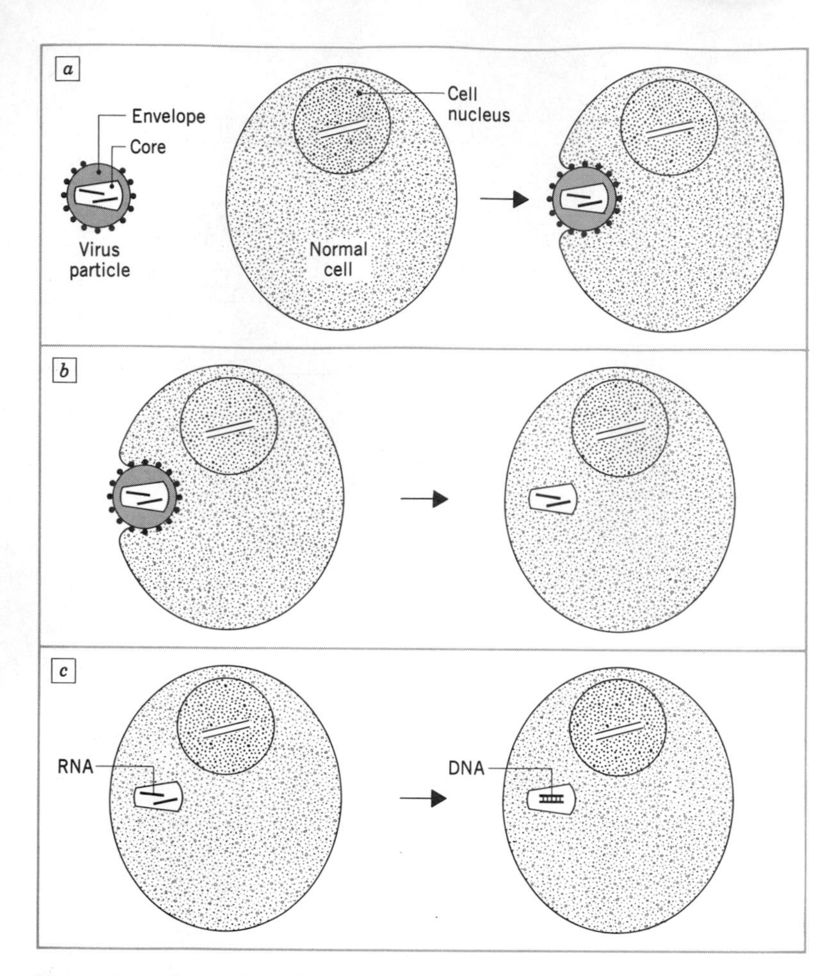

Figure 11-4 Retrovirus Life Cycle. (**a**) Attachment of virus to an animal cell. Specific proteins on the virus surface are thought to interact with specific receptors on the cell surface, leading to binding of the virus to the cell. (**b**) Viral penetration of an animal cell. The viral membrane fuses to the animal cell membrane, releasing the viral core into the cytoplasm. (**c**) Reverse transcription. Viral RNA is converted into DNA by reverse transcriptase. (**d**) Viral integration. Viral DNA, probably still packaged in the core, migrates to the nucleus of the cell and inserts into an animal chromosome. (**e**) Viral gene expression. RNA is transcribed from the integrated form of the virus. Transcription begins at the strong promoter in the left LTR and continues into the right LTR. This RNA is used to transcribe viral proteins, and for some this re-

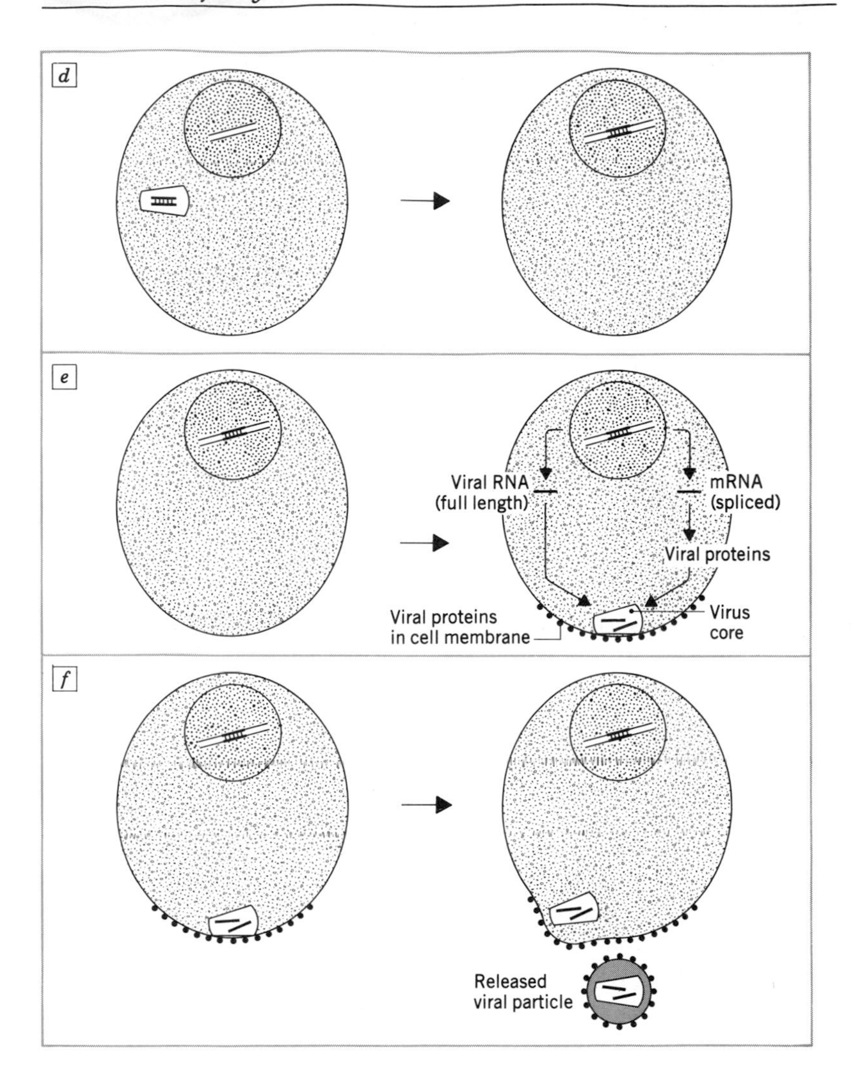

quires an additional splicing step. Some of the full-length RNA molecules are packaged into cores by viral proteins. Viral proteins from the envelope gene insert into the cell membrane of the host cell. **(f)** Virus release. Viral cores containing RNA form near the inner side of the cell membrane. They then appear to be surrounded by the cell membrane, which now contains viral surface proteins. In a sense the viral cores bud out of the cell, covered by cell membrane. Some of these budding steps can be seen in Figure 11-1*a*.

reverse transcription and integration are described in more detail below. Messenger RNA is later synthesized from the integrated virus (Figure 11-4*e*). Some of this RNA serves as the genetic information in new virus particles. Other RNA molecules are translated into viral proteins, and for some genes RNA splicing occurs. Viral proteins and full-length viral RNAs then spontaneously assemble into viral cores. The envelope proteins appear to be incorporated into the membrane of the cell, and as the viral cores leave the cell, they are wrapped by a protective coat derived from the cell membrane (Figure 11-4*f*; also see viruses budding from cells in Figure 11-1*a*). With most retroviruses, the infective process does not kill the cell; HIV-1 is an exception.

Reverse transcription occurs inside the viral core, after the core has entered the host cell. Most of the research effort on reverse transcription has focused on two problems: (1) How do the U5 and U3 regions, which are at opposite ends of the RNA, come to lie next to each other in the LTRs of the DNA? and (2) How is DNA synthesis primed? (DNA polymerases require a primer to begin synthesis; see Chapter 5, especially Figure 5-11). One possible scheme is sketched in Figure 11-5. As pointed out in the preceding section, the viral RNA contains a repeated sequence R at each end, as well as subterminal regions called U5 and U3 (see Figure 11-3*a*). There is a region just beyond the 3′ end of U5 that is complementary to a transfer RNA, and it appears that formation of a tRNA – retroviral RNA hybrid generates the primer necessary for reverse transcriptase to begin DNA synthesis (see Figure 11-5*a*). Reverse transcription then proceeds toward the 5′ end of the viral RNA, stopping when the end is reached. Reverse transcriptase contains an activity called **ribonuclease H** (RNase H), which degrades RNA hybridized to DNA; thus the ribonuclease H activity allows the protein to destroy the viral RNA once it has been used as a template, leaving only the DNA copy of the viral genetic information. Removal of the R region of the RNA frees the new DNA strand to hybridize with the R region of the second RNA molecule in the viral core. This may explain why there are two copies of RNA in the viral core. This strand transfer (Figure 11-5*d*) creates a new primer that can be used to continue synthesis to the end of the second RNA.

Next, the second strand of the DNA must be synthesized. This synthesis is thought to be primed by a short **oligonucleotide**, which binds just outside the U3 region. DNA synthesis from this primer results in a short piece of DNA that includes the site where the tRNA binds to

RNA, a location called the primer binding site. Circularization could lead to a second strand transfer and the generation of two more primers. DNA synthesis from these primers would then complete the process.

After the viral genetic information has been converted to a DNA form, it integrates (inserts) into the host DNA by a process that has features resembling bacterial transposition and the integration of lysogenic bacteriophages. Mutations at the end of either LTR or within the integrase gene block integration; thus these regions are crucial for integration. It is thought that the ends of the LTRs serve as binding sites for the integrase protein, and after insertion they are located at the ends of the integrated **provirus** (Figure 11-6). Integration occurs at a wide variety of places in the host DNA, and at each integration site a few base pairs of host DNA are duplicated. Thus there is a direct repeat of host DNA at the junctions of host and viral DNA. In its integrated form the virus can remain dormant for many years.

AIDS

Almost all people diagnosed with AIDS possess antibodies directed against HIV-1. This, plus the appearance of AIDS following transfusions with HIV-contaminated blood, first suggested that HIV-1 is responsible for the disease. The presence of the antibodies now serves as a test for people exposed to the disease. Many of these "seropositive" people show no symptoms of the disease, but it is anticipated that all or almost all will eventually exhibit a frank case of AIDS. During the early stages of AIDS infection there is often abundant replication of HIV-1, producing free virus in several parts of the body. This first wave of viral attack is often accompanied by fevers, rashes, and flu-like symptoms. But within a few weeks, the virus almost disappears, as do the symptoms. Several years later, the virus again rapidly replicates, and the final stages of infection set in.

A major cellular target of HIV-1 is a type of blood cell called a helper T cell. This type of cell normally participates in the immunological response of animals against bacteria, viruses and other microorganisms. Helper T cells called CD4 lymphocytes have on their surface the protein called CD4 or T4, which is also the receptor for HIV-1. Thus they are susceptible to infection and killing by HIV-1. AIDS patients have

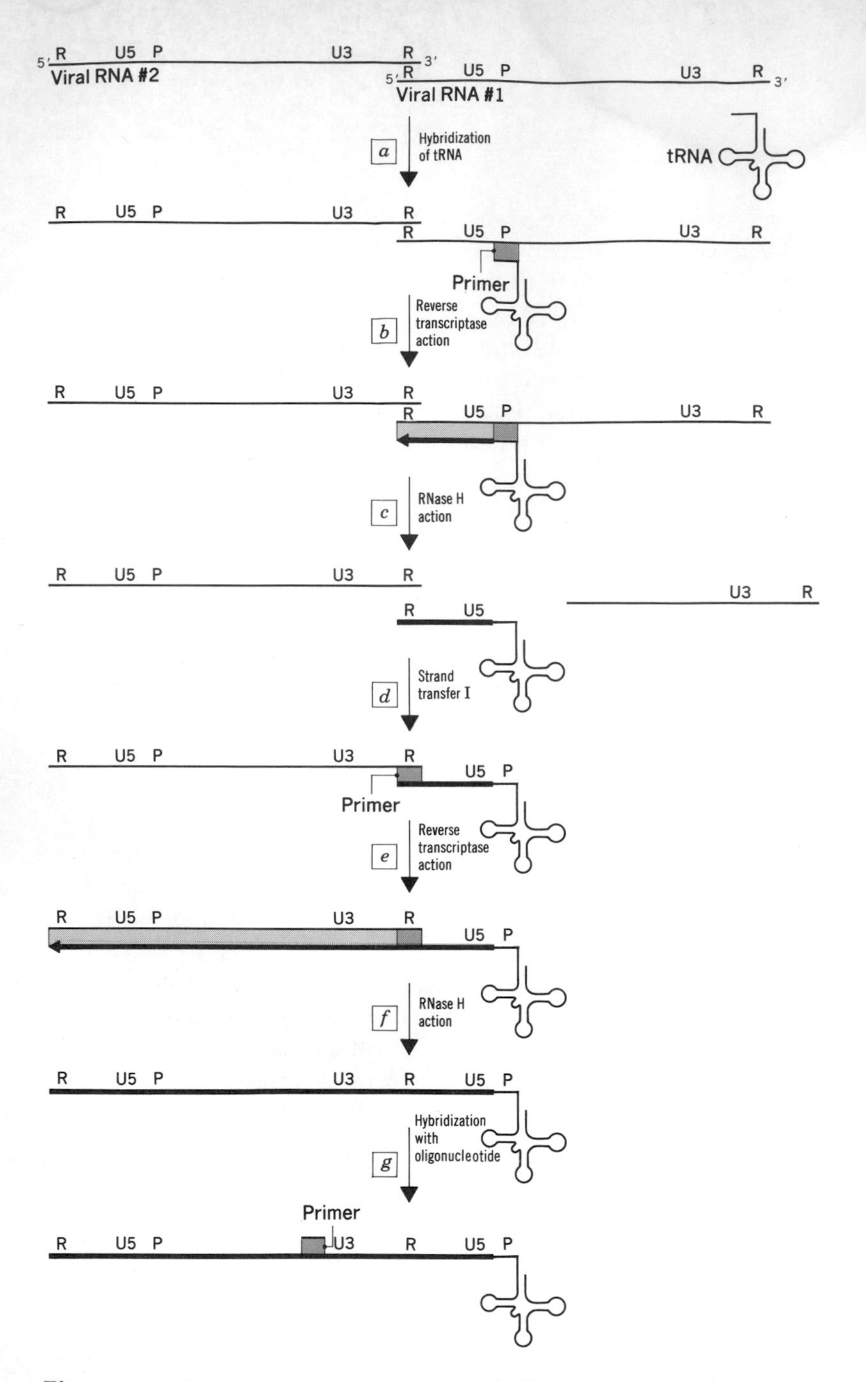

Figure 11-5 Synthesis of DNA from RNA. Reverse transcription converts the genetic information in two viral RNA molecules into a single DNA molecule. In the process regions labeled U5 and U3 are duplicated, and with region R they form the direct long terminal repeats called LTRs. One way this might happen is shown. The thin lines indicate RNA and the heavy ones DNA. Base pairing is indicated by hatch marks. (**a**) The viral RNA and a specific transfer RNA hybridize to create a primer on the RNA close to U5 at the site labeled **P**.

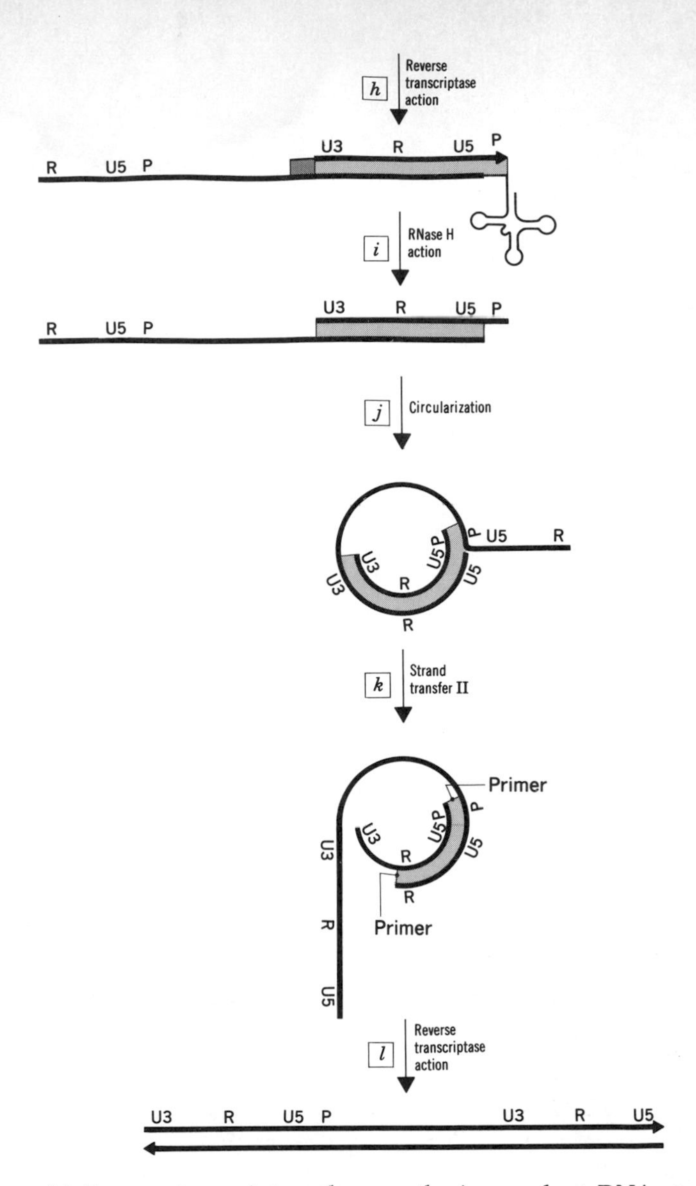

(**b**) Reverse transcriptase then synthesizes a short DNA strand from this primer. (**c**) RNase H action degrades regions of RNA hybridized to DNA. (**d**) Strand transfer occurs by hybridization of **R** regions of RNA and DNA. (**e**) Reverse transcription starts again, this time from the primer created by the 3′ end of the DNA. This generates one LTR (U3-R-U5). (**f**) The RNase H activity of the reverse transcriptase degrades the hybridized RNA, leaving only the DNA copy plus the tRNA. (**g**) An oligonucleotide hybridizes to a region adjacent to U3, forming a primer. (**h**) Reverse transcriptase action synthesizes DNA including the primer binding site (**P**). (**i**) RNase H action again degrades RNA hybridized to DNA, releasing the tRNA primer. (**j**) The DNA circularizes through hybridization of the regions labeled **P**. (**k**) A second strand transfer occurs to generate two primers. (**l**) Reverse transcription synthesizes the remainder of each strand.

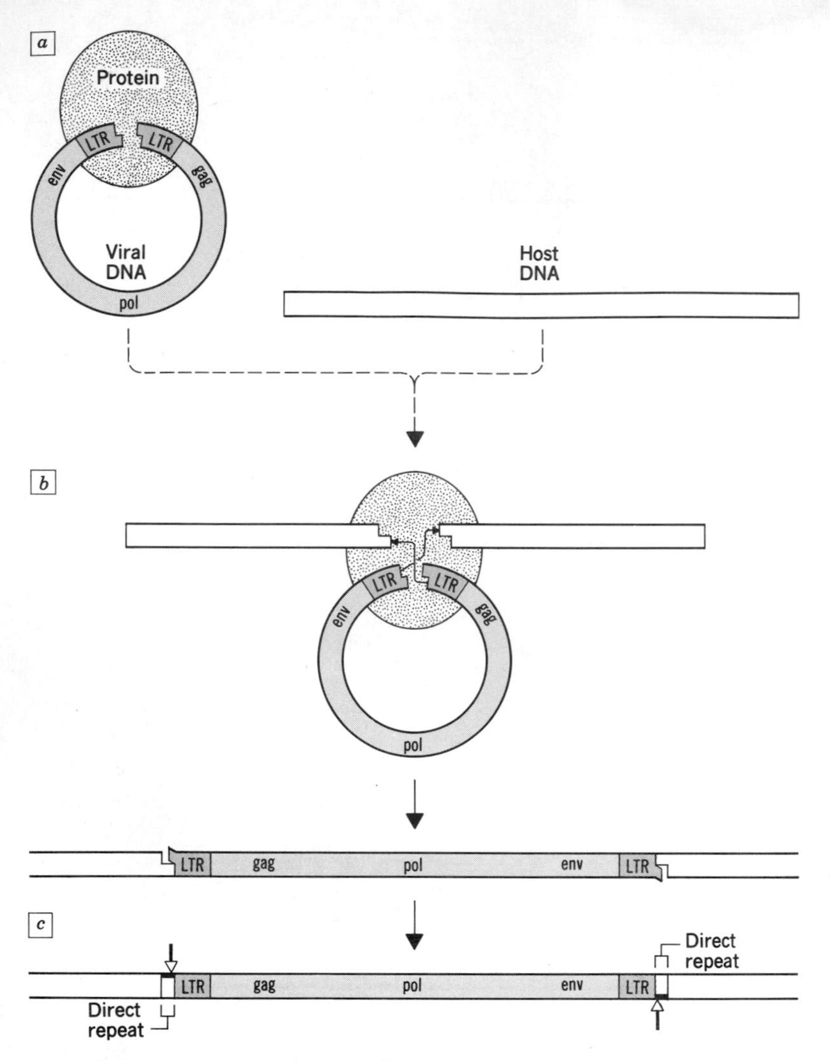

Figure 11-6 Integration of Retroviral DNA. (**a**) Retroviral DNA is present in the cell nucleus in a form that can react with chromosomal DNA. Although both circular and linear viral forms of DNA are present, it is the linear DNA that integrates. Following reverse transcription, integrase trims two base pairs from each 3′ end of the viral DNA. (**b**) A protein-mediated complex forms between viral and host DNA, opening the host DNA with a staggered cut (there are five to nine overhanging bases, depending on the virus). The viral DNA strands are ligated to the host DNA, as shown by the arrows. (**c**) The two bases overhanging at each end of the virus are removed, and the single-stranded regions are used as templates for DNA polymerase. DNA synthesis fills the gaps. Thus the inserted virus is bounded by direct repeats and has lost two base pairs from each end.

an abnormally low number of these cells, presumably because of the HIV-1 infection. The loss of these cells leaves the patient susceptible to a variety of organisms that normally are unable to establish infection.

HIV-1 has genetic and biochemical properties similar to those of other retroviruses but with additional features. Examination of the nucleotide sequence of the viral genome reveals six regions (*vif, vpr, vpu, tat, rev,* and *nef*: see Figure 11-3*b*) that are not found in most other retroviruses. These regions all encode proteins, several of which cause specific antibody production by AIDS patients. The Tat protein causes a 1000–2000-fold increase in transcription originating from the promoter located in the left LTR region. Since transcription of RNA for all of the viral genes begins at this promoter, Tat increases expression of all the viral genes. Viruses containing a mutation in *tat* show little infectivity. The Rev protein is also a positive regulator. It selectively favors production of proteins that will eventually become a part of the virus. Nef appears to be a negative regulator, inhibiting transcription of full-length viral RNA. The interplay of these regulators is probably important in the ability of the virus to grow rapidly or slowly at different stages of the infection cycle. The product of the *vif* gene appears to be involved in the ability of the virus to infect cells but not in viral replication. At present little can be said about *vpr* and *vpu.*

Although it is not completely clear why T4 lymphocytes are selectively killed by HIV-1, the viral envelope protein appears to be a key player through its interaction with the high concentration of CD4 receptors present on these cells. One scenario is that during infection the viral envelope protein accumulates on the surface of infected cells, binds to receptors on nearby uninfected cells, and effectively tears holes in their membranes. This would cause a cell to leak to death. Another type of cell death arises from the fusion of an infected cell with nearby cells, a process mediated by the envelope protein and the CD4 cellular receptor. A single infected cell can fuse with as many as 500 uninfected cells to form a giant cellular aggregate that eventually dies. A third process of killing is carried out by the immune system itself. Infected cells normally shed large amounts of the envelope protein (gp120), and in addition to stimulating antibody production, the envelope protein probably binds to CD4 receptors on uninfected cells. Consequently, host antibodies against gp120 would bind to these cells. Then other components of the immune system would recognize these uninfected, antibody-covered cells and kill them. Thus only one in a thousand T4 lymphocytes may actually be infected, but the viral envelope protein causes the whole population to be devastated.

ONCOGENES

When tissues in animals reach maturity, their constituent cells generally stop growing and dividing. As pointed out above, a comparable phenomenon occurs with cells grown in culture—normal cells of most types stop dividing when they have completely covered the surface of a petri dish. Tumor viruses perturb the biochemical pathways that control the growth and division of cells, resulting in uncontrolled cellular growth. Consequently, when cultured cells are exposed to a tumor virus, the infected cells continue growing and dividing even when the surface of the petri dish is covered. The infected cells grow on top of each other, forming clumps (**foci**) that can be easily observed using a low power microscope. In many respects foci are like bacterial colonies, and they are often used for the same type of manipulation. When transplanted into an animal, the infected cells will often develop into tumors.

Retroviruses are thought to cause tumors in two general ways (Figure 11-7). First, viral integration can occur near or within an important control gene in the host chromosome, raising the level of expression of that gene in a way that leads to uncontrolled cellular growth. In general, the probability of this happening is low because viral integration is not specifically targeted for these sites. Second, the virus itself may carry a gene whose product disrupts normal cellular control pathways. These virally borne "control genes" are called oncogenes, and they tend to produce tumor cells at a high frequency. Viral oncogenes are often defective copies of cellular regulatory genes. Oncogenes frequently interfere with the normal functioning of the viral genes responsible for producing new virus. Thus tumor viruses are often defective and sometimes require "helper" viruses to reproduce. This may help explain why epidemics of cancer have not been observed. To date about 50 different oncogenes have been identified.

One important class of oncogene is called *ras.* The *ras* genes are a small family of related genes found normally in a variety of eukaryotic organisms ranging from yeast to humans. They are associated with a variety of human and rodent tumors, and *ras* genes have been found as oncogenes in mouse and rat retroviruses. Nucleotide sequence comparisons between normal *ras* genes and oncogenic ones in retroviruses indicate that oncogenic *ras* genes contain mutations. Thus it may be that some **carcinogens** cause cancer by chemically modifying the *ras*

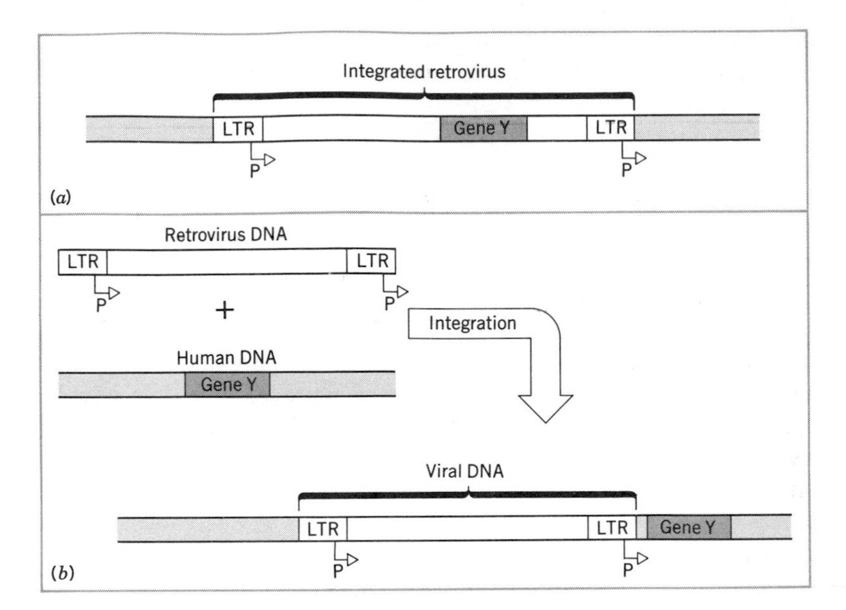

Figure 11-7 Stimulation of Gene Expression by Retroviruses. (a) Insertion of a retrovirus immediately upstream from a human gene (gene *Y*) can place that gene under the control of the strong promoter in the right LTR of the virus. Shaded regions indicate human DNA. **(b)** The viral DNA contains a copy of an animal gene (gene *Y*), which is now controlled by the strong viral promoter in the left LTR instead of by a normal cellular promoter. Shaded regions indicate human DNA.

region of the DNA. Normal *ras* genes can also produce malignancy if the Ras proteins are produced in very large amounts. This can occur if a very strong promoter is placed immediately upstream from the gene. Retroviruses contain strong promoters in their LTRs, and the one in the 3′ LTR stimulates transcription outside the integrated virus (Figure 11-7). Thus integration of the virus immediately upstream from a *ras* gene can lead to uncontrolled cell growth.

PERSPECTIVE

AIDS has brought retroviruses to almost everyone's attention. The disease was first clinically described in 1981. Six years later more than

50,000 Americans had been diagnosed with this condition, many of whom have died. Another million were thought to be infected by the virus, and the number keeps climbing. It is now clear that transmission is via body fluids and that the disease is generally spread by sexual contact or exchange of blood and blood products. The disease cannot be cured by antibiotics, since these drugs work on bacteria, not viruses. Infection does, however, require several proteins unique to the virus, proteins such as reverse transcriptase, integrase, a specific protease, and the products of the viral regulatory genes. An attractive goal is to design chemical agents that attack these proteins but have minimal effects on the host cells and, hopefully, few side effects.

Formation of cancer cells still is a complex process; we do not even understand how normal cell division is controlled. The identification of oncogenes such as *ras* should be helpful in studying the development of malignancy. But even though 5–40% of human tumors contain *ras* oncogenes, we cannot yet say the *ras* genes cause cancer because the appropriate control experiments cannot be done. Nevertheless, the nucleotide sequence comparisons may reveal certain mutant alleles of *ras* and other gene families that are associated with a predisposition to cancer. Then gene cloning technologies may become useful for identifying individuals whose genes contain the mutations so that such persons can be guided away from some high-risk activities involving known carcinogens.

As the research effort intensifies to learn about the AIDS virus, our understanding of retroviruses in general will deepen considerably. This knowledge should in turn help us treat certain types of cancer. But in the long run, the major impact of this research may be in the area of human engineering rather than AIDS, for with understanding of retroviruses comes the ability to use defective ones as cloning vehicles to insert genes into human chromosomes.

Questions for Discussion

1. How is AIDS spread?
2. How are the life cycles of bacteriophage lambda and the Human Immunodeficiency Virus similar and how are they dissimilar?

3. Methods are being devised to block the replication of retroviruses to stop the spread of AIDS. Some, such as the use of AZT, block reverse transcription, and others are aimed at inactivating regulatory proteins such as Tat. Based on what you know about the life cycle of HIV-1, are these approaches likely to cure someone of AIDS (i.e., to rid the person of the virus)?
4. If you were designing methods to find chemicals that would block stages in the virus life cycle, would you seek compounds that acted before or after integration?
5. What are some of the problems involved in using retroviruses as vehicles for inserting genes into humans?
6. How could a ribozyme be designed to protect cells from attack by HIV-1?

ADDITIONAL READING

Note: Older articles often have the most appropriate illustrations for nonscientists.

Ada, G. L., and Nossal, S. G. The Clonal-Selection Theory. *Scientific American*, August 1987, pp. 62–69.

This paper presents a historical perspective of the development of our understanding of the immune system.

Anderson, W. F., and Diacumakos, E. G. Genetic Engineering in Mammalian Cells. *Scientific American*, July 1981, pp. 106–121.

Strategies are described for using bacteria and recombinant DNA techniques to engineer mammalian cells.

Atkinson, M. A., and Maclaren, N. K. What Causes Diabetes? *Scientific American*, July 1990, pp. 62–71.

The relationship of the immune system to diabetes is discussed.

Bishop, J. M. Oncogenes. *Scientific American*, March 1982, pp. 80–92.

Oncogenes cause cancer. They were first reported in viruses, but they have also been found in normal cells. Abnormal expression of these genes can lead to cancerous growth.

Bretscher, M. S. How Animal Cells Move. *Scientific American*, December 1987, pp. 72–90.

With the exception of sperm, animal cells move by bringing into cytoplasm pieces of the cell membrane that are later recycled to the surface.

Brierley, C. L. Microbiological Mining. *Scientific American*, August 1982, pp. 44–53.

The use of a bacterium called *Thiobacillus* in leaching copper from low-grade ore is discussed.

Capra, J. D., and Edmundson, A. B. The Antibody Combining Site. *Scientific American*, January 1977, pp. 50–59.

Antibody structure is discussed, and the biochemical details of antigen-antibody binding are described.

Cech, T. R. RNA as an Enzyme. *Scientific American*, November 1986, pp. 64–75.

Self-processing of RNA is described, a concept that has led to the development of ribozymes.

Chambon, P. Split Genes. *Scientific American,* May 1981, pp. 60–71.
Some of the experimental evidence leading to the concept of introns and exons is presented.

Chilton, M.-D. A Vector for Introducing New Genes into Plants. *Scientific American,* June 1983, pp. 51–59.
Some plant tumors are caused by a bacterium that can be used to introduce genes into plants.

Collier, R. J., and Kaplan, D. A. Immunotoxins. *Scientific American,* July 1984, pp. 56–71.
Chemical modification of antibodies may lead to the development of a class of antitumor agents.

Croce, C. M., and Klein, G. Chromosome Translocations and Human Cancer. *Scientific American,* March 1985, pp. 54–60.
The movement of oncogenes and their activation can be traced by chromosome mapping methods.

Darnell, J. E. The Processing of RNA. *Scientific American,* October 1983, pp. 90–101.
In higher cells RNA is modified in a number of ways before it reaches the ribosomes in the cell cytoplasm.

Darnell, J. E. RNA. *Scientific American,* October 1985, pp. 68–87.
The various roles played by RNA are described.

DeRobertis, E. M. Oliver, G., and Wright, C. V. E. Homeobox Genes and the Vertebrate Body Plan. *Scientific American,* July 1990, pp. 46–52.
The homeobox genes are similar in many animals, and their involvement in the control of development of limbs and organs is discussed.

Dickerson, R. E. The DNA Helix and How It Is Read. *Scientific American,* December 1983, pp. 94–111.
This paper presents a detailed treatment of DNA structures.

Ding, E., Young, J., and Cohn, Z. A. How Killer Cells Kill. *Scientific American,* January 1988, pp. 38–45.
Killer lymphocytes are the commandos of the immune system. They secrete protein molecules that punch holes in cells, causing them to leak to death.

Donelson, J. E., and Turner, M. J. How the Trypanosome Changes Its Coat. *Scientific American,* February 1985, pp. 44–51.
The causative agent of sleeping sickness evades the host immune response by switching on new genes that encode the surface protein of the parasite.

Doolittle, R. F. Proteins. *Scientific American,* October 1985, pp. 88–99.
Protein structure and enzyme action are described.

Edelman, G. Topobiology. *Scientific American,* May 1989, pp. 76–88.
Topobiology, the study of place-dependent interactions, may be revealing information about the origin of the immune system.

Fedoroff, N. V. Transposable Genetic Elements in Maize. *Scientific American,* June 1984, pp. 84–98.

An introduction to mobile genetic elements in plants.

Felsenfeld, G. DNA. *Scientific American,* October 1985, pp. 58–67.

DNA is flexible, and along its length its structure can change in ways that are probably important for gene control.

Freifelder, D. (editor). *Recombinant* DNA. W. H. Freeman & Company, San Francisco, 1978, 147 pp.

A collection of articles from *Scientific American* that provides background material on the development of recombinant DNA technology.

Gallo, R. C. The AIDS Virus. *Scientific American,* January 1987, pp. 46–73.

This, the following paper by the same author, and the paper by Hazeltine and Wong-Staal lay a foundation for understanding the AIDS virus.

Gallo, R. C. The First Human Retrovirus. *Scientific American,* December 1986, pp. 88–101.

This lays part of the foundation for understanding the AIDS virus.

Geever, R. F., Wilson, L. B., Nallaseth, F. S., Milner, P. F., Bittner, M., and Wilson, J. T. Direct Identification of Sickle-Cell Anemia by Blot Hybridization. *Proceedings of the National Academy of Sciences (U.S.),* 78:5081–5085 (1981).

This research paper describes how sickle-cell anemia can be diagnosed.

Gilbert, W., and Villa-Komaroff, L. Useful Proteins from Recombinant Bacteria. *Scientific American,* April 1980, pp. 74–94.

Recombinant DNA techniques are described with special reference to the procedures used to clone an insulin gene.

Godson, G. N. Molecular Approaches to Malaria Vaccines. *Scientific American,* May 1985, pp. 52–59.

The proteins of the outer coat of the malaria parasite appear to serve as decoys to deflect the host immune system.

Grey, H., Sette, A., and Buus, S. How T Cells See Antigen. *Scientific American,* November 1989, pp. 56–64.

T cells are an important part of the immune system; their mode of interaction with other cells of the immune system is described.

Grivell, L. A., Mitochondrial DNA. *Scientific American,* March 1983, pp. 78–89.

Mitochondria are subcellular organelles in which chemical energy is converted into a useful form (ATP). These organelles have their own genetic system.

Grobstein, C. *A Double Image of the Double Helix.* W. H. Freeman & Company, San Francisco, 1979, 177 pp.

During the mid-1970s recombinant DNA technologies led to public controversies. This book discusses the controversies and provides a copy of the NIH guidelines that regulate recombinant DNA research.

Hakomori, S. Glycosphingolipids. *Scientific American,* May 1986, pp. 44–53.

These fat molecules are parts of cell membranes. Cell membrane structure is described, along with changes that occur at the onset of cancer.

Hazeltine, W., and Wong-Staal, F. The Molecular Biology of the AIDS Virus. *Scientific American,* October 1988, pp. 52–62.

This is one of 10 papers on AIDS in the October 1988 issue of *Scientific American.*

Hirsch, M. S., and Kaplan, J. C. Antiviral Therapy. *Scientific American,* April 1987, pp. 76–85.

Viruses are not susceptible to the antibiotics we normally use to kill bacteria. Strategies are described for obtaining antiviral agents.

Holliday, R. A Different Kind of Inheritance. *Scientific American,* June 1989, pp. 60–73.

The methylation of DNA may be an important way in which gene activity patterns are passed from one generation to another.

Hunter, T. The Proteins of Oncogenes. *Scientific American,* August 1984, pp. 70–79.

A discussion of ways in which oncogenes may cause cancer.

Jonathon, P., Butler, G., and Klug, A. The Assembly of a Virus. *Scientific American,* November 1978, pp. 62–69.

This description of the assembly of Tobacco Mosaic Virus provides an introduction into the details of virus structure.

Kennedy, R. C., Melnick, J. L., and Dreesman, G. R. Anti-Idiotypes and Immunity. *Scientific American,* July 1986, pp. 48–69.

This is a discussion of antibodies that recognize other antibodies, a process that is probably important in the modulation of the normal immune system.

Kornberg, R. D., and Klug, A. The Nucleosome. *Scientific American,* February 1981, pp. 52–64.

Higher cells package their DNA by wrapping it around ball-like structures made of protein.

Lake, J. A. The Ribosome. *Scientific American,* August 1981, pp. 84–97.

A three-dimensional model is presented in this description of how proteins are made.

Laurence, J. The Immune System in AIDS. *Scientific American,* December 1985, pp. 84–93.

A description of T4 lymphocytes, one of the cell types attacked by the AIDS virus.

Lawn, R. M., and Vehar, G. A. The Molecular Genetics of Hemophilia. *Scientific American,* March 1986, pp. 48–65.

Hemophilia is caused by a defective gene. The product of the corresponding normal gene has now been produced artificially.

Leder, P. The Genetics of Antibody Diversity. *Scientific American,* May 1982, pp. 102–115.

The shuffling of segments of DNA and RNA that occurs during the formation of antibody genes is considered.

Maniatis, T., Hardison, R. C., Lacy, E., Lauer, J., O'Connell, C., Quon, D., Sim, G., and Efstratiadis, A. The Isolation of Structural Genes from Libraries of Eucaryotic DNA. *Cell,* **15**:687–701 (1978).

This research paper describes the cloning of rabbit hemoglobin genes.

Marrack, P., and Kappler, J. The T Cell and Its Receptor. *Scientific American,* February 1986, pp. 36–45.

T cells are an important factor in the ability of the immune system to react specifically to viruses. This article introduces the many factors involved in an immune response.

Matthews, T., and Bolognesi, D. AIDS Vaccines. *Scientific American,* October 1988, pp. 120–127.

This is a good treatment of the immune response and vaccine development.

McKnight, S. L. Molecular Zippers in Gene Regulation. *Scientific American,* April 1991, pp. 54–64.

The interaction of polypeptide subunits of a protein is discussed.

Mills, J., and Masur, H. AIDS-Related Infections. *Scientific American,* August 1990, pp. 50–57.

Opportunistic infections are the major killers in AIDS.

Mullis, K. The Unusual Origin of the Polymerase Chain Reaction. *Scientific American,* April 1990, pp. 56–65.

PCR is a very powerful way to amplify specific regions of a DNA molecule. This is a personal story of the discovery of PCR.

Murray, A. W., and Kirschner, M. W. What Controls the Cell Cycle? *Scientific American,* March 1991, pp. 56–63.

A protein has been discovered that seems to be a regulator of the cell cycle in many organisms.

Murray, A. W., and Szostak, J. W. Artificial Chromosomes. *Scientific American,* November 1987, pp. 62–87.

The behavior of whole chromosomes can be studied by creating artificial chromosomes using genetic engineering.

Nathans, J. The Genes for Color Vision. *Scientific American,* February 1989, pp. 42–49.

A molecular explanation for color blindness is beginning to emerge.

Neufeld, P., and Coleman, N. When Science Takes the Witness Stand. *Scientific American,* May 1990, pp. 46–53.

Some of the pros and cons of DNA fingerprinting are discussed.

Nomura, M. The Control of Ribosome Synthesis. *Scientific American,* January 1984, pp. 102–115.

A description of ribosomes and the genes that encode their components.

Pestka, S. The Purification and Manufacture of Human Interferons. *Scientific American,* August 1983, pp. 36–43.

Human cells release a protein called interferon, which protects against infection by some viruses. Its manufacture by gene cloning techniques is described.

Ptashne, M. How Gene Activators Work. *Scientific American,* January 1989, pp. 40–47.

The interaction of specific proteins with DNA can lead to transcription of certain genes, and considerable detail is known about these interactions.

Ptashne, M., Johnson, A. D., and Pabo, C. O. A Genetic Switch in a Bacterial Virus. *Scientific American,* November 1982, pp. 128–140.

When bacteriophage lambda infects a cell, it can either kill its host or coexist with it. This article describes how the virus makes the choice.

Radman, M., and Wagner, R. The High Fidelity of DNA Duplication. *Scientific American,* August 1988, pp. 40–46.

This article describes how three enzymes work together to synthesize DNA, to proofread, and to correct mistakes.

Rennie, J. The Body Against Itself. *Scientific American,* December 1990, pp. 106–115.

The immune system is discussed with particular attention on autoimmune diseases.

Richards, F. M. The Protein Folding Problem. *Scientific American,* January 1991, pp. 54–63.

To be active proteins must fold in a very precise way. Some ideas about folding and protein structure are discussed.

Ross, J. The Turnover of Messenger RNA. *Scientific American,* April 1989, pp. 48–55.

Genes can be controlled by way of timing the breakdown of messenger RNA.

Sachs, L. Growth, Differentiation and the Reversal of Malignancy. *Scientific American,* January 1986, pp. 40–47.

Growth and differentiation are described with a focus on leukemia.

Sapienza, C. Parental Imprinting of Genes. *Scientific American,* October 1990, pp. 52–60.

Identical genes can have different effects depending upon whether they came from the father or the mother.

Scheller, R. H., and Axel, R. How Genes Control an Innate Behavior. *Scientific American,* March 1984, pp. 54–83.

Egg-laying behavior in a marine snail is coupled to the expression of specific genes.

Simons, K., Garoff, H., and Helenius, A. How an Animal Virus Gets Into and Out of its Host Cell. *Scientific American,* February 1982, pp. 58–66.

The virus causes the host cell to manufacture new virus particles, including a protective membrane that originates from the host cell membrane.

Smith, K. Interleukin-2. *Scientific American,* March 1990, pp. 50–57.

Interleukin-2 is thought to be a hormone that helps control the immune system.

Stahl, F. W. Genetic Recombination. *Scientific American,* February 1987, pp. 90–101.

DNA molecules naturally break and rejoin with other molecules to give new combinations of nucleotide sequences.

Steitz, J. Snurps. *Scientific American,* June 1988, pp. 56–63.

Snurps are small nuclear ribonucleoproteins involved in the removal of introns from RNA in higher cells.

Tiollais, P., and Buendia, M. A. Hepatitis B Virus. *Scientific American,* April 1991, pp. 116–123.

This deadly virus is being attacked by recombinant DNA techniques.

Tonegawa, S. The Molecules of the Immune System. *Scientific American,* October 1985, pp. 122–131.

This article presents a detailed description of antibodies and the genetic rearrangements that lead to antibody diversity.

Unwin, N. The Structure of Proteins in Biological Membranes. *Scientific American,* February 1984, pp. 78–95.

The discussion includes an explanation of how protein structure is studied.

Varmus, H. Reverse Transcription. *Scientific American,* September 1987, pp. 56–65.

The conversion of RNA to DNA is a central step in the infection of cells by retroviruses such as the AIDS virus. But this process also occurs in a number of other biological systems, and it may even reflect a very early process in the evolution of genetic material.

Verma, I. Gene Therapy. *Scientific American,* November 1990, pp. 68–84.

The feasibility of introducing healthy genes into people to correct diseased genes is discussed.

Wang, J. C. 1982 DNA Topoisomerases. *Scientific American,* July 1982, pp. 94–109.

A class of enzyme is described that is able to convert rings of DNA from one topological form to another. These enzymes are probably important in many aspects of chromosome function.

Watkins, P. Restriction Fragment Length Polymorphism (RFLP): Applications in Human Chromosome Mapping and Genetic Disease Research. *Biotechniques,* **6**:310–320 (1988).

Watson, J. D. *The Double Helix* (Gunther Stent, Ed.) W. W. Norton & Company, New York, 1980. 298 pp.

Watson's account of the discovery of DNA structure is placed in historical perspective by Gunther Stent, Francis Crick, Linus Pauling, and Aaron Klug. Stent has also collected background materials, which include reproductions of original scientific papers and reviews of Watson's story by well-known scientists.

Watson, J. D., and Tooze, J. *The DNA Story.* W. H. Freeman & Company, San Francisco, 1981, 605 pp.

Included in this documentary history of gene cloning are reproductions of press clippings and personal letters written by scientists involved. The section titled "Scientific Background" provides a historical introduction to gene cloning.

Weinberg, R. A. A Molecular Basis of Cancer. *Scientific American,* November 1983, pp. 126–143.

The relationship of oncogenes and cancer cells is discussed.

Weinberg, R. A. The Molecules of Life. *Scientific American,* October 1985, pp. 48–57.

This is an overview of DNA, proteins, and gene cloning.

Weintraub, H. Antisense RNA and DNA. *Scientific American,* January 1990, pp. 40–46.

Antisense oligonucleotides are capable of blocking gene expression, and they have potential as therapeutic agents.

White, R., and Lalouel, J. Chromosome Mapping with DNA Markers. *Scientific American,* February 1988, pp. 40–48.

This, and the article by Watkins, describe how restriction fragment length polymorphisms are used to study the genetics of disease.

APPENDIX ONE

TRANSMISSIBLE PLASMIDS

Moving Genes

Transmissible plasmids contain genes whose protein products allow them to naturally go from one bacterial cell to another. The process of plasmid movement is called **conjugation**. The best studied case of conjugation involves a plasmid of *E. coli* called **F** or **fertility factor**. The plasmid DNA contains genes coding for proteins responsible for the formation of long, filamentlike structures that protrude from the outside of the bacterial cell. Such a protrusion is called a **pilus; pili** are thought to be important in attaching or attracting plasmid-containing cells to plasmid-deficient ones. When the two types of cells get close

---

Figure AI-1 Bacterial Conjugation. (**a**) An *E. coli* cell containing an F plasmid forms long, flexible tubular structures (pili) on its surface (1–3 per cell). Each pilus is composed of proteins encoded by genes located on the F plasmid. (**b**) One pilus binds to an *E. coli* cell that lacks an F plasmid. (**c**) The pilus retracts, pulling the two cells close to each other. (**d**) A break occurs in one of the DNA strands of the F plasmid; one of the ends of the broken strand rolls off the circular strand and passes into the recipient cell. (**e, f**) Soon after the single strand has reached the interior of the recipient cell, DNA polymerase (not shown) begins to make a complementary strand. At about the same time, a new copy of the transferred strand is made in the donor cell. (**g**) Both donor and recipient cells now have a complete copy of the F plasmid. (**h**) The two cells separate. Each contains a copy of the F plasmid. The recipient cells form pili. Now both can act as donor cells.

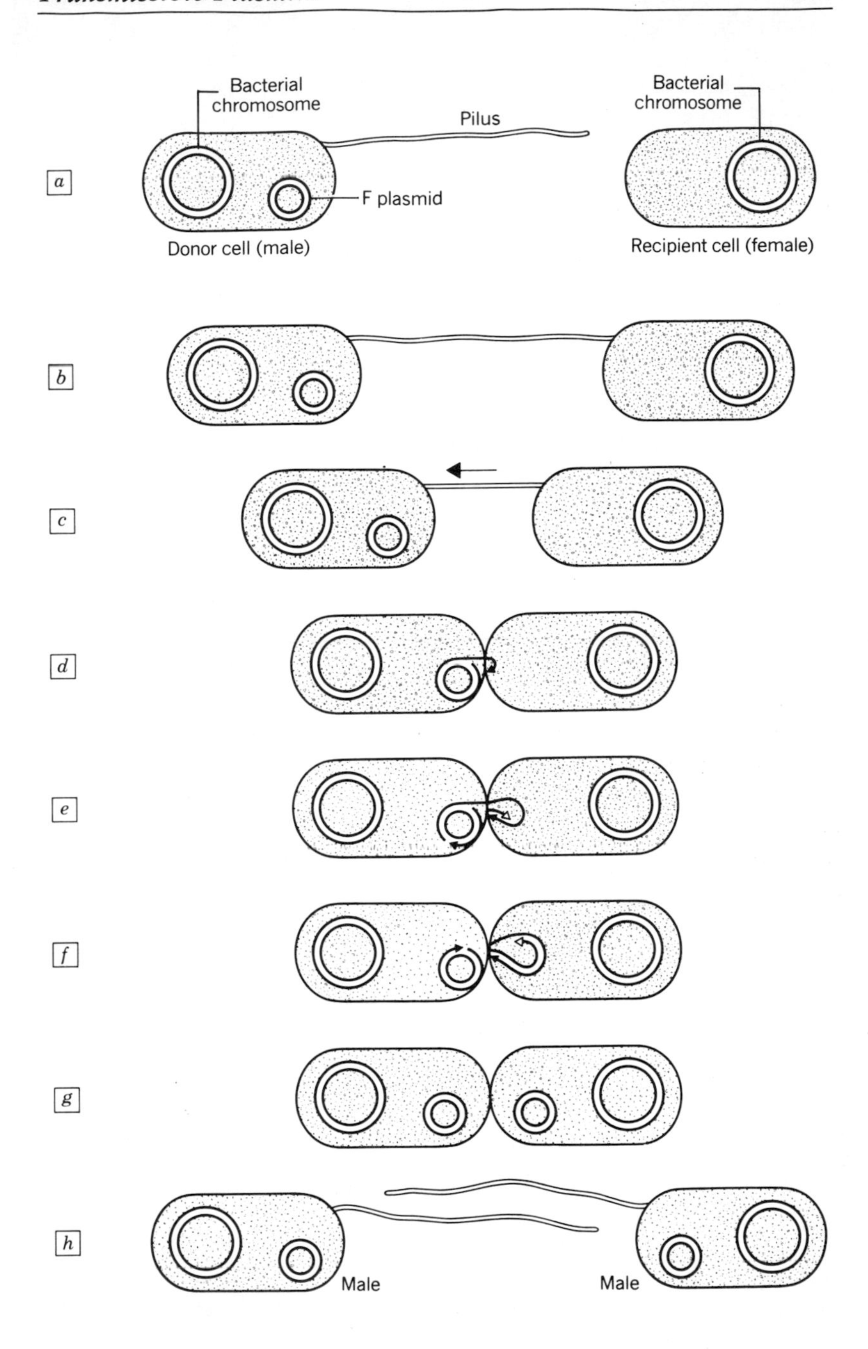
a
Bacterial chromosome
Pilus
Bacterial chromosome
F plasmid
Donor cell (male)
Recipient cell (female)
b
c
d
e
f
g
h
Male
Male

enough, they mate. One strand of the plasmid DNA is transferred from the plasmid-containing cell to the plasmid-deficient one. As this occurs, both strands are used as templates to synthesize new DNA. Thus both cells of the mating pair end up infected with a plasmid (Figure AI-1). On rare occasions the plasmid DNA and the bacterial chromosomal DNA can join to form a giant circle. When this happens, genes in the bacterial chromosome can be transferred from one cell to another by the plasmid mating process. In a sense, bacterial conjugation is a primitive form of sex.

In general, transmissible plasmids are not considered to be good cloning vehicles because they could conceivably lead to accidents. For example, most laboratory strains of *E. coli* are useful as tools because they contain a number of known genetic defects. As a result of the defects, these weakened organisms are not expected to survive well outside the laboratory, and in general, the laboratory strains seem unlikely to cause infections. However, if genes were to be cloned into laboratory bacteria by using transmissible plasmids, and if the "engineered" bacteria containing the plasmids and cloned genes were to accidentally escape from the laboratory, the plasmids and the cloned genes could conceivably be transferred, through the process of conjugation, into a healthy, wild strain of the bacterium. The wild strain might have a greater chance than the weak, laboratory strain to enter human digestive tracts and thus place an unwanted gene where it could do harm. Fortunately there are a large number of *non*transmissible plasmids of varying sizes and degrees of complexity, so it is unnecessary to use transmissible plasmids for gene cloning.

A P P E N D I X T W O

WALKING ALONG DNA

Cloning Adjacent Fragments of DNA

Once a piece of DNA has been cloned, determining the sequence is straightforward. Often molecular biologists want to know the nucleotide sequence of adjacent regions of DNA. This information is obtained by a modification of the cloning procedure called **walking along DNA**. The general strategy is to use a nucleotide sequence near one end of the cloned region as a probe for locating adjacent, overlapping regions in a collection of recombinant DNAs (Figure AII-1). Two restriction endonucleases are used. First, one is found that cuts at or near an end of the cloned fragment (X in Figure AII-1*a*). Then another is found that cuts at a site in the cloned sequence and also at a site far outside the cloned region (R in Figure AII-1*a*). When human DNA is cut with enzyme R only and the fragments are cloned into plasmids, one type of recombinant DNA (region II, Figure AII-1*a*) will partially overlap with the original cloned sequence. The overlapping sequence is indicated as a solid region in Figure AII-1. Bacterial colonies containing this DNA are identified by the colony hybridization technique (Figure 8-5) using a radioactive probe from the overlapping region obtained as outlined in Figure AII-1. The DNA fragment (region II, Figure AII-1) can be isolated from recombinant plasmids by methods involving gel electrophoresis, and the nucleotide sequence of this region can be determined. By repeating this process, one can move along a DNA molecule, successively cloning and sequencing small bits of DNA.

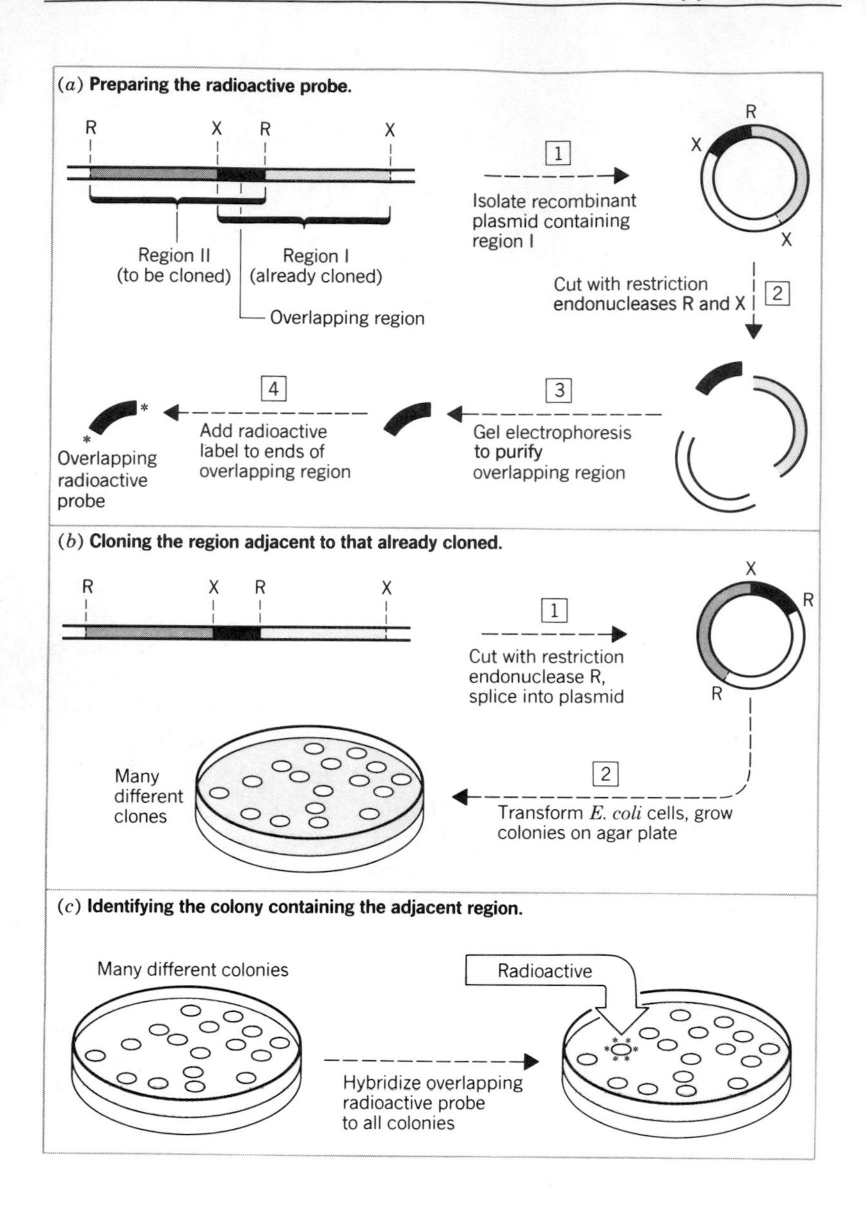
(a) Preparing the radioactive probe.
R
X
R
X
Region II
(to be cloned)
Region I
(already cloned)
Overlapping region
1
Isolate recombinant plasmid containing region I
Cut with restriction endonucleases R and X
2
3
Gel electrophoresis to purify overlapping region
4
Add radioactive label to ends of overlapping region
Overlapping radioactive probe
(b) Cloning the region adjacent to that already cloned.
Cut with restriction endonuclease R, splice into plasmid
Transform E. coli cells, grow colonies on agar plate
Many different clones
(c) Identifying the colony containing the adjacent region.
Many different colonies
Hybridize overlapping radioactive probe to all colonies
Radioactive

Figure AII-1 Walking Along DNA. (**a**) Preparing the radioactive probe. A short piece of DNA is shown with two adjacent, partially overlapping regions. Region I has already been cloned and is bounded by cleavage sites for restriction endonuclease X. Region II is to be cloned and is bounded by cleavage sites for restriction endonuclease R. (**1**) Region I DNA is inserted into a plasmid. (**2**) Region I–plasmid recombinant DNA is cut by endonuclease R and X to liberate the short piece of DNA (solid) that represents the overlap between regions I and II. (**3**) The overlapping DNA piece is separated from all other pieces by gel electrophoresis (see Figure 7-4). (**4**) Radioactive label is enzymatically attached to the overlapping fragment. (**b**) Cloning the region adjacent to that already cloned. (**1**) Total DNA is cut with restriction endonuclease R and is inserted into plasmid DNA, producing recombinant DNA molecules of many types. (Only the particular recombinant DNA being sought is illustrated. In this DNA the region of overlap is present because endonuclease X has not been used.) (**2**) *E. coli* cells are transformed with the recombinant plasmids, and these cells are grown into colonies on agar plates. Very few of these colonies contain the adjacent region (region II). (**c**) Identifying the colony containing the adjacent region. The many colonies obtained in (b) are tested by nucleic acid hybridization using the overlapping radioactive probe prepared in (a). The colony containing region II will become radioactive, and it can be identified by exposure of X-ray film (see Figure 8-5). All plasmid DNA isolated from this colony will contain region II.

APPENDIX THREE

MONOCLONAL ANTIBODIES

Tools for Manipulating Proteins

When we are exposed to viruses, bacteria, or other antigens (foreign macromolecules), our immune system produces antibodies that bind to the antigens. Chapter 10 discussed how DNA rearrangements can lead to the formation of billions of different types of antibody, giving our immune system the capacity to recognize billions of different antigens. These rearrangements and other differentiation events give rise to a large population of cells called B cells, each of which recognizes a different **antigenic determinant**, that is, a different section or structural aspect of a protein or other large molecule. Binding of a B cell to an antigen stimulates that B cell to multiply and differentiate, leading to a large population of cells that produce antibodies against the particular antigen.

Since most large molecules, such as proteins, have a number of different antigenic determinants, a particular antigen generally stimulates many different B cell clones to multiply. All of these produce antibodies that bind to the particular antigen, but each B-cell clone produces antibody that binds to only one particular region of the antigen. This is called a **polyclonal** immune response. Methods have been developed to grow clones from single B cells, clones that produce large quantities of a single type of antibody. Such antibodies, called **monoclonal** antibodies, are rapidly becoming valuable research and clinical tools because they are so specific.

Efficient laboratory production of monoclonal antibodies involves the fusion of two different types of cell. First, an animal is immunized with the antigen against which one wishes to produce antibodies. This is generally done by several injections of the antigen over a period of weeks or months. A collection of B cells is then obtained, usually by dissecting the animal's spleen. The B cells are mixed with cultured **myeloma cells** under conditions in which the two cell types fuse to form what is called a **hybridoma**. Myeloma cells are malignant (cancerous) antibody-producing cells. When fused to a B cell, a myeloma cell stimulates the B cell to secrete large amounts of the antibody that the B-cell had been programmed to produce. The myeloma cell also immortalizes the B cell (cultured B cells normally die within a few days after removal from the body).

Of course, the fusion mixture contains many uninteresting cell types and fusion products. To allow only the hybridomas to grow in culture, genetic defects have been introduced into the myeloma cells. The strategy is similar to the use of antibiotic resistance in gene cloning to obtain a particular desired bacterial colony. Once the culture has been enriched for hybridomas, single cells are cultured separately and allowed to multiply. The resulting cultures are then tested for the production of the desired antibody. In this procedure, myeloma variants that have stopped producing their own antibodies are used; thus, the resulting hybridomas produce only the antibody made by the B cell.

Specific antibody–protein interactions play the same role in the study of protein biology that nucleic acid hybridization plays in the analysis of genes. Antibodies can be used to separate specific proteins from other macromolecules, to determine the location of a protein in the cell, to quantify various proteins in body fluids, and even to clone genes. Monoclonal antibodies greatly enhance the precision of analysis because the antibody preparation recognizes only a single determinant on the protein. Moreover, once cultured, the cells producing the monoclonal antibodies always produce the same antibody; this is not true when animals are used as antibody sources, since their antibody specificities often change over time.

In clinical medicine the potential use of monoclonal antibodies ranges from diagnosis of diseases to therapy. Their exquisite specificity makes antibodies well-suited for identifying cells of specific types. For example, antibodies that recognize tumor cells could be used to locate tumors and then evaluate the effectiveness of surgery or chemotherapy. Small cytotoxic molecules could even be attached to the antibodies; when the antibodies were injected into cancer victims, they would attach to the cancer cells and kill them. Infectious diseases could also be fought with monoclonal antibodies by injection either before or soon after exposure to dangerous viruses and bacteria. Indeed, this form of passive immunization is one of the approaches being taken to try to combat AIDS.

GLOSSARY

adenine one of the bases that forms a part of DNA or RNA. It is usually abbreviated by the letter A.

agar a gelatinlike substance obtained from seaweed. Used in petri dishes to culture bacteria, agar allows microbiologists to obtain bacterial colonies.

agar plate a petri dish containing solid agar.

amino acid a building block of protein. The 20 different amino acids have a common structure shown below. The letter R represents chemical side chains, which are different for each amino acid. The chemical properties of the side chains help determine how a protein folds, so the arrangement of amino acids dictates the three-dimensional structure of a protein.

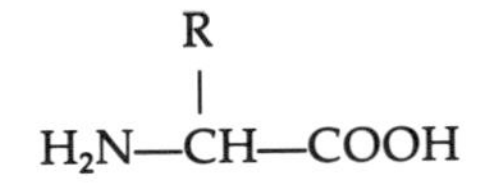

amino acids	
alanine (ala)	leucine (leu)
arginine (arg)	lysine (lys)
asparagine (asn)	methionine (met)
aspartic acid (asp)	phenylalanine (phe)
cysteine (cys)	proline (pro)
glutamine (gln)	serine (ser)
glutamic acid (glu)	threonine (thr)
glycine (gly)	tryptophan (trp)
histidine (his)	tyrosine (tyr)
isoleucine (ile)	valine (val)

aminoacyl-tRNA synthetases members of a class of enzymes that link specific amino acids with specific transfer RNA molecules. One synthetase recognizes one particular type of transfer RNA and one particular type of amino acid.

ampicillin an antibiotic related to penicillin. Both these drugs kill bacteria by preventing cell wall synthesis.

antibiotic a substance that kills an organism. In the present context an antibiotic is a drug that kills bacteria. Common examples are streptomycin, erythromycin, penicillin, ampicillin, and tetracycline.

antibiotic-resistance gene a gene that codes for a protein, which then allows a bacterium to live in the presence of a drug that normally would kill it. Some resistance genes change the target of the drug so it no longer binds the drug, others cause active secretion of the drug, and still others break down the drug. Plasmids often contain such genes.

antibodies proteins that recognize and bind to antigens. Antibodies are an important component of the immune system.

anticodon a particular three-nucleotide region in transfer RNA that is complementary to a specific three-nucleotide codon in messenger RNA. Alignment of codons and anticodons is the basis for ordering amino acids in a protein chain.

antigen a foreign chemical or microorganism that is recognized by, and attaches to, an antibody.

antigenic determinant a particular region of an antigen that is recognized by a particular antibody. Most antigens have many determinants, and so when an immune response develops, many different types of antibodies will be present.

antisense RNA an RNA molecule that is the complement of another RNA molecule. The two RNA molecules have the ability to form a hybrid. Antisense RNA molecules can be designed to hybridize with particular mRNA molecules and thereby prevent the mRNA from being translated.

assay a method or way of measuring chemical compounds.

atom a particle composed of a nucleus (protons and neutrons) and electrons. Common atoms are carbon, oxygen, nitrogen, and hydrogen. These atoms differ from one another by having different numbers of protons, neutrons, and electrons. Groups of atoms bonded together produce molecules.

ATP adenosine triphosphate, a relatively small molecule that serves as an energy carrier and as one of the precursors to RNA. ATP has high energy bonds that are easily broken by enzymes to release the energy needed to drive some of the cellular chemical reactions.

attenuation a type of gene control in which a signal present in the messenger RNA can stop RNA polymerase from making the messenger RNA longer. The signal is activated by high concentrations of the product of the reactions controlled by the genes being transcribed into the messenger RNA.

B lymphocyte a type of cell in mammals that produces antibodies.

bacterial culture a batch of bacterial cells grown either on solid agar or in a brothlike solution.

bacteriophage a virus that attacks bacteria; also called a phage.

bacterium (plural: bacteria) a one-celled organism lacking a nucleus, mitochondria, and chloroplasts. Although many biochemical properties of bacteria differ from those of higher organisms, the basic features of chemical reactions are very similar in bacteria and man.

base (1) A flat ring structure, containing nitrogen, carbon, oxygen, and hydrogen, that forms part of one of the nucleotide links of a nucleic acid chain. The bases are adenine, thymine, guanine, cytosine, and uracil, commonly abbreviated A, T, G, C, and U. (2) A hydrogen ion acceptor, like sodium hydroxide (lye).

base pair (bp) two bases, one in each strand of a double-stranded nucleic acid molecule, which are opposite each other. The bases of a base pair are attracted to each other by weak chemical interactions. Only certain pairs form: A-T, G-C, and A-U. See Figure 3-5.

broth a liquid culture medium used to grow bacteria. One common type contains yeast extract, beef extract, table salt, and water.

capsid the external protein coat of a virus particle.

carcinogen a chemical that causes cancer, generally by altering the structure of DNA (see **mutagen**).

cell the smallest unit of living matter capable of self-perpetuation; an organized set of chemical reactions capable of self-production. A cell is bounded by a membrane that separates the inside of the cell from the outer environment. Cells contain DNA, where information is stored, ribosomes, where proteins are made, and mechanisms for converting energy from one form to another.

cell extract or lysate a mixture of cellular components obtained by mechanically or enzymatically breaking cells. The cell extract is the starting material from which biochemists obtain enzymes, RNA, and DNA.

cell wall a thick, rigid structure surrounding cells of certain types, especially bacterial and plant cells. Cell walls are often composed of complex sugars.

cellulose the principal molecule in paper and wood; chemically, cellulose is a very large molecule composed of repeating units of glucose, a six-carbon sugar.

centrifuge a machine used to create gravitylike forces. Centrifuges can create forces hundreds of thousands of times that of gravity, making it possible to quickly separate molecules on the basis of size and shape. Merry-go-rounds and the spin cycle mechanisms of automatic clothes washing machines are examples of centrifuges.

centrifuge rotor the part of a centrifuge that holds test tubes and rotates at high speed.

chemical reaction a rearrangement of atoms to produce a set of molecules that are different from the starting molecules.

chromosome a subcellular structure containing a long, discrete piece of DNA plus the proteins that organize and compact the DNA.

clone (1) noun: a group of identical cells, all derived from a single ancestor.

(2) verb: to perform or undergo the process of creating a group of identical cells or identical DNA molecules derived from a single ancestor.

cloning vehicles small plasmid, phage, or animal virus DNA molecules into which a DNA fragment is inserted so the fragment can be transferred from a test tube into a living cell. Cloning vehicles are capable of multiplying inside living cells. Thus, if a cloning vehicle transfers a specific fragment of DNA into a cell that is also multiplying, all the progeny of that cell will contain identical copies of the vehicle and the transferred DNA fragment.

codon three nucleotides whose precise order corresponds to one of the 20 amino acids. In addition, special codons that do not code for any amino acid act as stop signals. In some cases several different codons encode the same amino acid.

cointegrate a type of DNA molecule thought to be an intermediate in transposition. Cointegrates contain the donor DNA, the recipient DNA, and two copies of the transposon.

colony a visible cluster of bacteria on a solid surface. All members of the colony arose from a single parental cell, and the colony is considered to be a clone. All members are identical. A colony generally contains millions of individual cells.

column chromatography a process for separating one type of molecule from another. A mixture of molecules is passed through a glass tube containing a material that binds the different molecular types with varying degrees of tightness (see Figure 5-9).

complementary describing two objects having shapes that allow them to fit together very closely: plugs and sockets, locks and keys, A's and T's or U's, G's and C's.

complementary base pairing rule Only certain nucleotides can align opposite each other in the two strands of DNA: G pairs with C; A pairs with T (or U in RNA).

complementary DNA (cDNA) DNA synthesized from RNA in test tubes using an enzyme called reverse transcriptase. The DNA sequence is thus complementary to that of the RNA. Complementary DNA is usually made with radioactive nucleotides and is used as a hybridization probe to detect specific RNA or DNA molecules.

conjugation a plasmid-mediated process of pairing between two bacteria with the transfer of genetic information (DNA) from the plasmid-containing bacterium to the plasmid-lacking bacterium.

constant region a region of an antibody molecule that is identical from one antibody molecule to another in a given class of antibody.

cytosine one of the bases that forms a part of DNA or RNA. It is usually abbreviated with the letter C.

denature to unfold, to become inactive. In reference to DNA, denaturation means conversion of double-stranded DNA into single-stranded DNA. In reference to proteins, denaturation means unfolding of the protein.

density gradient a solution in which the density gradually increases from top to bottom. Density gradients are commonly used to help separate large molecules in a centrifuge.

deoxyribonucleic acid see **DNA**.

dissolve to disperse a solid substance in a liquid.

DNA deoxyribonucleic acid. DNA is a long, thin, chainlike molecule that is usually found as two complementary chains and is often hundreds to thousands of times longer than the cell in which it resides. The links or subunits of DNA are the four nucleotides called adenylate, cytidylate, thymidylate, and guanylate. The precise arrangement of these four subunits, repeated many times, is used to store all the information necessary for life.

DNA ligase the enzyme that joins two separate DNA molecules together end to end.

DNA polymerase the enzyme complex that makes new DNA using the information contained in old DNA.

DNA replication the process of making DNA. DNA is always made from preexisting information in DNA (or, in special cases, from RNA). DNA replication involves a number of different enzymes.

E. coli *Escherichia coli.* These bacteria are commonly found in the digestive tracts of many mammals including humans.

egg (1) germ cell produced by a female. (2) an animal embryo, along with a food supply, enclosed by a shell or membrane.

electron micrograph a photograph taken using an electron microscope, an instrument that is similar to a light microscope but uses a beam of electrons to expose the film rather than a beam of light. Because the effective wavelength of electrons is much shorter than that of light, objects that are measured in millionths of a centimeter can be seen using an electron microscope.

element one of slightly more than 100 distinct types of matter, which singly or in combination compose all materials of the universe. An atom is the smallest representative unit of an element.

embryo a plant or animal in an early stage of development, generally still contained in a seed, egg, or uterus.

encode contain a nucleotide sequence specifying that one or more specific amino acids be incorporated into a protein.

endonuclease an enzyme that cuts DNA or RNA at points inside the molecule (i.e., away from the ends).

enhancer a short region of DNA that influences an activity of DNA from a distance. Most enhancers that have been studied stimulate expression of genes that are located far from the enhancers.

envelope a covering or coat; the outermost coat of an animal virus.

enzyme a protein molecule, or occasionally an RNA molecule, specialized to catalyze (accelerate) a biological chemical reaction. Generally enzyme names end in *-ase.*

equilibrium the absence of *net* movement one way or another.

expression see **gene expression**.

expression vector a plasmid or phage in which the DNA contains an active, but easily controlled promoter, downstream from which a gene of interest can be inserted. Following induction of the promoter, the protein of interest can be produced in large amounts, sometimes up to 40% of the total cellular protein of the bacterium that carries the vector.

F the abbreviation for fertility factor; a type of transmissible plasmid.

Fertility factor F, a type of transmissible plasmid found in *E. coli.* that can cause transfer of chromosomal DNA from one cell to another.

fetus (adj **fetal**) an embryo in a late stage of development, but still in the uterus.

fission a type of cell division in which a parental cell divides in half to form two daughter cells.

foci (singular: **focus**) clumps of animal cells that have been stimulated by virus infection to continue to grow and divide when uninfected cells in the culture stop dividing. The number of foci is related to the number of virus particles used to infect the culture, so the effective virus concentration can be measured by counting foci.

frameshift displacement of the nucleotide reading frame in DNA or RNA. Frameshifts are generally caused by addition or deletion of one or more nucleotides. Since the nucleotides are read in units of three, addition or deletion of three nucleotides (or multiples of three) will have no effect on the reading frame.

gel electrophoresis a method for separating molecules based on their size and electric charge. Molecules are forced to run through a gel (e.g., gelatin) by placing them in an electric field. The speed at which they move depends on their size and charge. See Figure 7-4.

gene a small section of DNA that contains information for construction of one protein molecule or in special cases for construction of transfer RNA or ribosomal RNA.

gene cloning a way to use microorganisms to produce millions of identical copies of a specific region of DNA.

gene expression the process of making the product of a gene. This involves transferring information, via messenger RNA, from a specific region of DNA, a gene, to ribosomes where a specific protein is made.

genetic engineering the manipulation of the information content of an organism to alter the characteristics of that organism. Genetic engineering may use simple methods like selective breeding or complicated ones like gene cloning.

genome the primary repository of genetic information for an organism; generally refers to the DNA molecule(s) and its nucleotide sequence. The bacterial genome is a single DNA molecule, while the human genome consists of DNA molecules from each of the chromosomes. In certain viruses, genetic information is stored in an RNA form.

genomic clone a DNA fragment derived directly from cellular DNA rather than from messenger RNA, the usual source for cDNA clones. Genomic and cDNA clones have different sequences due to RNA splicing.

germ cells a particular type of cell responsible for creating the next generation; also called gametes. In most higher organisms body cells contain two sets of chromosomes; germ cells contain only one set. Thus when two germ cells join together, the resulting cell (zygote) has two sets of chromosomes. This cell then produces new body (somatic) cells.

globin one of the protein chains that comprise hemoglobin.

guanine one of the bases that forms a part of DNA or RNA. It is usually abbreviated with the letter G.

heavy chain the larger of the two types of protein chain that comprise an antibody.

hemoglobin the blood protein responsible for transporting oxygen to the tissues.

heteroduplex mapping a process of locating regions in one DNA molecule that are homologous to those in another. Two DNAs are denatured and allowed to form hybrids. Homologous regions become double-stranded while nonhomologous regions remain single-stranded. The hybrids are examined by electron microscopy, and the lengths of the double-stranded and single-stranded regions are measured.

histones small, DNA-binding proteins involved in the packaging of DNA; the four "core" histones form a structure around which about 200 base pairs of DNA are wrapped.

homologous corresponding or similar in position; describing regions of DNA molecules that have the same nucleotide sequence. Since DNA has two complementary strands, complementary base pairing can occur between homologous regions in two different DNA molecules. Homologous also refers to regions of DNA, RNA or protein that are similar due to a common ancestry.

host an organism that provides the life support system for another organism, virus, or plasmid. *E. coli* is a host for certain plasmids that exist inside the bacterium, and we are the host for *E. coli,* for these bacteria live inside us.

hybrid a double-stranded nucleic acid in which the two complementary strands differ in origin.

hybridoma the hybrid product of a fusion between a myeloma cell and another cell.

hydrogen bond a weak, attractive force in which a hydrogen atom of one molecule is drawn toward another molecule.

hypervariable region a short section of amino acids in an antibody molecule that frequently exhibits differences in amino acid sequence from one antibody molecule to another.

induce cause to happen, often with reference to gene expression. Specific molecules called inducers bind to certain repressors and prevent the repressor from binding to DNA.

infectious capable of invading a host.

insulin a protein involved in the control of sugar metabolism in mammals. Insulin is made by cells of the pancreas.

integrase a protein involved in the insertion of one DNA molecule into another.

integration the insertion of one DNA molecule into another.

interferon a protein made in the body that helps fight virus infections.

intron a noncoding section of a gene that is removed from RNA before translation in cells from higher organisms. Bacterial messenger RNA does not contain introns.

kinase an enzyme that adds a phosphate to another molecule.

lambda (λ) the name of a particular bacteriophage used extensively in gene cloning.

leader a region of film, RNA, or protein that precedes the region of primary information content. In RNA the leader extends from the first nucleotide at the 5′ end to the codon specifying the first amino acid of the protein. A protein leader, if present, is usually defined as a region at the amino terminal end that is cut from the protein during movement of the protein across membranes.

light chain the smaller of the two types of protein chain that comprise an antibody.

LTR abbreviation for long terminal repeat, the nucleotide sequence that is repeated at the ends of a retrovirus DNA.

lye sodium hydroxide (NaOH); caustic soda. Dilute lye solutions will cause DNA strands to separate, thus producing single-stranded DNA.

lysogen a bacterium harboring a phage that does not kill the bacterium and also protects it from further infection by other related phages. A phage called lambda (λ) often causes *E. coli* cells to become lysogens.

lysozyme an enzyme that breaks down bacterial cell walls. Lysozyme can be obtained from egg white or tears.

lytic infection a type of viral infection in which the cell being attacked dies and then disintegrates (lyses). In contrast, a lysogenic infection does not lead to cell death.

messenger RNA (mRNA) RNA used to transmit information from a gene on DNA to a ribosome, where the information is used to make protein.

metabolism a collective term for all the chemical reactions involved in life. For example, sugar metabolism includes the reactions that occur in the body during the production, use, and breakdown of sugars.

micrometer one millionth of a meter.

milligram 0.001 gram (28 grams = 1 ounce).

millimeter 0.001 meter (1 meter = about 39 inches).

missense a type of mutation in which a nucleotide change causes an incorrect amino acid to be incorporated into a protein.

mitochondrion (pl. **mitochondria**) a specialized subcellular structure that converts chemical energy from one form to another.

molecule a group of atoms tightly joined together. The arrangement of atoms is very specific for a given molecule, and this arrangement gives each molecule specific chemical and physical properties. The oxygen molecule we breathe is two oxygen atoms bonded together. Paper is largely cellulose molecules, which are giant molecules containing carbon, oxygen, and hydrogen.

monoclonal referring to a single type of clone or cell line; usually referring to a clone of cells that produces a single type of antibody, which recognizes only a single determinant on the antigen used to stimulate antibody production.

monolayer a layer of cells one cell thick.

mutagen an agent that increases the rate of mutation by causing changes in the nucleotide sequences of DNA (see **carcinogen**).

mutant an organism whose DNA has been changed relative to the DNA of the dominant members of the population.

mutations errors in DNA, often occurring during DNA replication, that cause incorrect amino acids to be inserted into proteins.

myeloma cell a type of tumor cell that produces antibodies.

nuclease a general term for an enzyme that cuts DNA or RNA.

nucleic acid DNA, RNA, or a DNA:RNA hybrid.

nucleic acid hybridization a process in which two single-stranded nucleic acids are allowed to base pair and form a double helix. The process makes it possible to use one nucleic acid to detect the presence of another having nucleotide sequence similarity. See Figure 5-10.

nucleoid a compact, DNA-containing structure found in bacteria; the bacterial equivalent of a chromosome.

nucleotide one of the building blocks of nucleic acids. A nucleotide is composed of three parts: a base, a sugar, and a phosphate. The sugar and the phosphate form the backbone of the nucleic acid, while the bases lie flat like steps of a staircase. DNA is composed of deoxyadenylate, deoxythymidylate, deoxyguanylate, and deoxycytidylate, four different kinds of nucleotide represented by the letters A, T, G, and C. See also **sequence**.

nucleotide pair two nucleotides, one in each strand of a double-stranded nucleic acid molecule, which are attracted to each other by weak chemical interactions between the bases. Only certain pairs form: A-T, G-C, and A-U.

nucleus (1) the core of an atom consisting of protons and neutrons; (2) a distinct subcellular structure containing chromosomes.

oligonucleotide a short piece of DNA or RNA containing three or more nucleotides. The oligonucleotides used in gene cloning are generally less than 100 nucleotides long, but in formal terms an oligonucleotide can be much longer.

operator a region on DNA capable of interacting with a repressor, thereby controlling the functioning of an adjacent gene.

operon a series of genes transcribed into a single RNA molecule. Operons allow coordinated control of a number of genes whose products have related functions.

organism one or more cells organized in such a way that the unit is capable of reproduction.

origin of replication a special nucleotide sequence that serves as a start signal for DNA replication.

pathogen a disease-causing agent (e.g., viruses that cause polio, mumps, and measles; bacteria that cause cholera and leprosy).

penicillin an antibiotic that kills *E. coli* and many other bacteria by blocking formation of new cell walls. Penicillin is produced by a mold.

peptide a short chain of amino acids, a fragment of a protein.

peptide bond the type of chemical bond that links two adjacent amino acids together in a protein chain.

phage bacteriophage.

phage plaques clear zones, created by bacteriophages killing bacteria, in a lawn of bacteria on an agar plate.

phenol an oily organic chemical used to separate DNA from proteins. Phenol is added to a mixture of DNA and proteins in water and the mixture is vigorously shaken. Proteins tend to move into the phenol; DNA stays in the water phase. The mixture is then briefly centrifuged. Phenol is more dense than water, so it forms a layer under the water. The water layer, containing the DNA, is removed with a pipette, leaving proteins behind in the phenol.

phosphate (PO_4) a chemical unit in which four oxygen atoms are joined to one phosphorus atom. The backbones of DNA and RNA are alternating phosphate and sugar units.

pilus (pl. **pili**) a long, hairlike structure produced on the surface of a bacterium containing a certain type of plasmid. Plasmid genes code for the proteins that comprise the pilus.

pipette a long, thin glass tube used for measuring volumes of liquids. A pipette can be used like an eyedropper to add or remove liquids from test tubes.

plaques see **phage plaques**.

plasmids small, circular DNA molecules found inside bacterial cells. Plasmids reproduce every time the bacterial cell reproduces.

point mutation a change of only one nucleotide pair in a DNA molecule.

poly A or **poly T tail** a long stretch (more than 20) of pure A's or T's, respectively, at the end of a DNA or RNA strand.

polyclonal referring to a mixture composed of many different clones; usually referring to an immune response in which there are many different antibodies that recognize different parts of an antigen.

polymerase chain reaction a test tube reaction in which a specific region of DNA is amplified many times by repeated synthesis of DNA using DNA polymerase and specific primers to define the ends of the amplified region.

polyprotein a long protein that is cleaved into several smaller proteins. The smaller proteins are thought to be the functional forms.

precipitate molecules that are clumped together so that they fail to pass through a filter. Precipitates are large aggregates and settle out of solution rapidly, much like silt out of river water.

prenatal before birth.

primer a piece of DNA or RNA that provides an end to which DNA polymerase can add nucleotides.

probe a DNA or RNA molecule, usually radioactive, which is used to locate a complementary RNA or DNA by hybridizing to it. Often a probe is used to identify bacterial colonies that contain cloned genes and to detect specific nucleic acids following separation by gel electrophoresis.

product the new molecules produced by a chemical reaction.

progeny offspring.

promoter a short nucleotide sequence on DNA where RNA polymerase binds and begins transcription.

protease an enzymatic protein that breaks down other proteins.

protein a class of long, chainlike molecules often containing hundreds of links called amino acids. There are 20 different amino acids used to make proteins. The thousands of different proteins serve many functions in the cell. As enzymes they control the rate of chemical reactions, and as structural elements they provide the cell with its shape. Members of this group of molecules are also involved in cell movement and in the formation of cell walls, membranes, and protective shells. Some proteins also help package the long DNA molecules into chromosomes.

protein synthesis see **translation**.

provirus a viral genome when integrated into a host chromosome.

pseudogene a nonfunctional gene that closely resembles a functional one.

purify to separate or isolate one type of molecule away from other types.

radioactive the state in which a substance (a molecule in the context of this book) contains an unstable element that spontaneously emits a high energy particle or radiation. The emission is detectable by photographic film, by Geiger counter, and by other instruments. Gene cloners generally use radioactive hydrogen, carbon, or phosphorus, all of which are commercially available. Radioactive uranium and plutonium are used in nuclear reactors.

radioactive tracer a radioactive atom that is incorporated into a specific molecule such that the molecule can be detected and measured by the presence of the radioactivity. It is much easier to measure small amounts of radioactivity than small amounts of particular chemicals.

reading frame one of three possible ways of reading a nucleotide sequence as a series of triplets such that the triplets correspond to the amino acids in the protein specified by the nucleic acid.

recognition site a short series of nucleotides specifically recognized by a protein, usually leading to the binding of that protein to the DNA at or near the point of the recognition sequence. Once the protein has bound to the DNA, it may cut, modify, or cover the DNA, depending on the function of the protein.

recombinant DNA molecule a DNA molecule containing two or more regions of different origin (e.g., plasmid DNA joined to a fragment of human DNA).

recombination the breaking and rejoining of DNA strands to produce new combinations of DNA molecules. Recombination is a natural process that generates genetic diversity. Specific proteins are involved in recombination.

replica plating the process in which colonies of bacteria are transferred from one agar plate to another without changing their relative orientation.

replication fork the point at which the two parental DNA strands separate during DNA replication.

repression a method of preventing gene expression in which a protein molecule (repressor) binds to the DNA near where RNA polymerase ordinarily would bind.

repressor a protein molecule that is capable of preventing transcription of a gene by binding to DNA in or near the gene.

restriction endonucleases enzymes that cut DNA at specific nucleotide sequences. The function of this class of enzyme inside cells is to protect the cells against invasion by foreign DNA.

restriction fragment length polymorphism a region of DNA that has several forms (the region differs from one individual to another) such that cleavage of DNA using a restriction endonuclease generates a DNA fragment whose size varies from one individual to another.

restriction mapping a procedure that uses restriction endonucleases to produce specific cuts in DNA. The positions of the cuts can be measured and oriented relative to each other to form a crude map.

retrovirus a type of animal virus whose life cycle involves conversion of genetic information from an RNA form to a DNA form.

reverse transcriptase an enzyme purified from tumor viruses that makes DNA from RNA.

RFLP restriction fragment length polymorphism.

ribonuclease H an enzyme that degrades RNA, but only when the RNA is in a hybrid double helix with a strand of DNA.

ribonucleic acid RNA.

ribosomes large, ball-like structures that act as workbenches where proteins are made. A bacterial ribosome consists of two balls, a small one called 30S and a larger one called 50S. Ribosomes are composed of special RNA molecules (ribosomal RNA) and about 50 specific proteins (ribosomal proteins).

ribozyme an RNA molecule that acts catalytically to cleave itself or another RNA molecule.

RNA ribonucleic acid. RNA is a long, thin, chainlike molecule usually found as a single chain. The links, or subunits, of RNA are the four nucleotides abbreviated as A, U, G, and C. Cells contain a number of different kinds of RNA molecule, which play different roles. The most common RNAs are messenger RNA, transfer RNA, and ribosomal RNA. Some viruses have RNA as their genetic material.

RNA:DNA hybrid a double-stranded molecule composed of one strand of RNA and one of DNA. The nucleotide sequences in the DNA and RNA are complementary.

RNA polymerase the enzyme complex responsible for making RNA from DNA. RNA polymerase binds at specific nucleotide sequences (promoters) in front of genes in DNA. It then moves through a gene and makes an RNA molecule that contains the information contained in the gene. Bacterial RNA polymerase makes RNA at a rate of about 65 nucleotides per second.

RNA splicing the process of removing regions from RNA. The removed regions are called introns, and the regions spliced together are called exons.

RNA tumor virus a type of RNA-containing virus that produces tumors in animals or converts normal cells in culture into tumor cells.

sequence the order of. In reference to DNA or RNA, sequence means the order of nucleotides.

sigma (σ) with reference to RNA polymerase, this Greek letter is used to signify a polypeptide (protein) subunit of the enzyme complex. The sigma subunit is important for promoter selection by RNA polymerase.

sodium hydroxide (NaOH) lye, caustic soda. A chemical that separates DNA strands.

somatic pertaining to the body. When referring to a type of cell, somatic means body cell rather than a sperm- or egg-producing cell.

Southern hybridization (Southern blotting) a method for transferring DNA from an agarose or acrylamide gel to nitrocellulose paper followed by hybridization to a radioactive probe.

sperm germ cell produced by a male.

sterile without life; generally referring to an instrument or a solution that has been heated to kill any organisms that may have been on or in it. Wire is sterilized by heating in a flame until it is red hot. Culture medium (e.g., broth) is sterilized by heating in a pressure cooker.

sticky ends specific termini (ends) of double-stranded DNA in which one of the strands sticks out farther than the other and the protruding strand is complementary to the protruding strand at the other end of the DNA molecule or at the end of another DNA molecule.

submicroscopic not visible when examined with a light microscope.

substrate the molecules on which an enzyme acts.

subunit one of the pieces that forms a part of a multicomponent structure, such as a link in a chain or a brick in a wall.

sugar a class of molecule containing particular combinations of carbon, hydrogen, and oxygen. The sugars in DNA and RNA are five-carbon sugars called deoxyribose and ribose, respectively. Glucose is a sugar containing six atoms of carbon per molecule.

sugar metabolism a group of biochemical reactions responsible for the formation of sugars and the conversion of sugars into other compounds.

tetracycline an antibiotic that kills bacteria by blocking protein synthesis.

tetramers four subunits, often identical. Many proteins are composed of separate polypeptide chains that act as subunits, associating to form the active protein.

3′ and 5′ ends the backbone of a nucleic acid molecule is composed of repeating phosphate and sugar subunits such that on one side of the sugar the phosphate is linked to the 5′ carbon of the sugar and on the other side the phosphate is linked to the 3′ carbon of the sugar (. . . phosphate-5′ carbon-4′ carbon-3′ carbon-phosphate-5′ carbon-4′ carbon-3′carbon . . .). When a chain is broken, the break generally occurs between the phosphate and the sugar. This produces two different ends. If only the sugar is considered, a 5′ carbon will be at one end (the 5′ end) and a 3′ carbon will be at the other (the 3′ end). These terminal carbons generally have a phosphate or an —OH group attached to them. See Figure 3-2.

thymine one of the bases that forms part of DNA. It is not found in RNA. It is usually abbreviated by the letter T.

topoisomerase an enzyme that breaks and rejoins DNA strands in such a way that it changes the number of times one strand crosses the other. Topoisomerases can tie and untie DNA knots, add and subtract twists in the DNA, and link and unlink DNA circles.

toxin a substance, often a protein in the context of this book, that causes damage to the cells of an organism.

transcription the process of converting information in DNA into information in RNA. Transcription involves making an RNA molecule using the information encoded in the DNA. RNA polymerase is the enzyme complex that executes this conversion of information.

transfer RNAs (tRNAs) small RNA molecules (each about 80 nucleotides long) that serve as adapters to position amino acids in the correct order during protein synthesis. The ordering by tRNA uses information in messenger RNA and occurs before the amino acids are linked together.

transformation the process whereby a bacterial cell takes up free DNA such that information in the free DNA becomes a permanent part of the bacterial cell. Often this means infecting a cell with a plasmid. With animal cells transformation means the conversion of a normal cell into a tumor cell.

translation the process of converting the information in messenger RNA into protein. Also called protein synthesis.

transposase a protein encoded by a gene in a transposon and required for transposition.

transposition the process whereby one region of DNA moves to another. Transposition often involves duplication of the region that moves.

tranposon a short section of DNA capable of moving to another DNA molecule or to another region of the same DNA molecule.

tryptophan one of the 20 amino acids found in proteins.

ultraviolet light a type of light that has very high energy and is invisible; black light. Nucleic acids absorb ultraviolet light, and instruments are available that measure the amount of absorption. The amount of absorption depends on the amount of nucleic acid present. Thus, by measuring the amount of absorption, it is possible to measure the amount of nucleic acid present.

uracil one of the bases that forms part of RNA. It is generally not found in DNA. It is usually abbreviated with the letter U.

variable region a region of an antibody molecule that differs in amino acid sequence from one antibody to another. Variable regions are thought to exist where *antigens* bind to antibodies.

vector see **expression vector**.

virus particles a class of infectious agents usually composed of DNA or RNA surrounded by a protective protein coat.

walking along DNA a gene cloning procedure used to clone regions of DNA adjacent to those already cloned.

yeast one-celled organism containing a true nucleus and mitochondria. Many biochemical properties of yeast are similar to those of higher organisms, and yeast are probably more closely related to mammals than to bacteria.

INDEX

FLOW OF GENETIC INFORMATION
replication, transcription, and translation

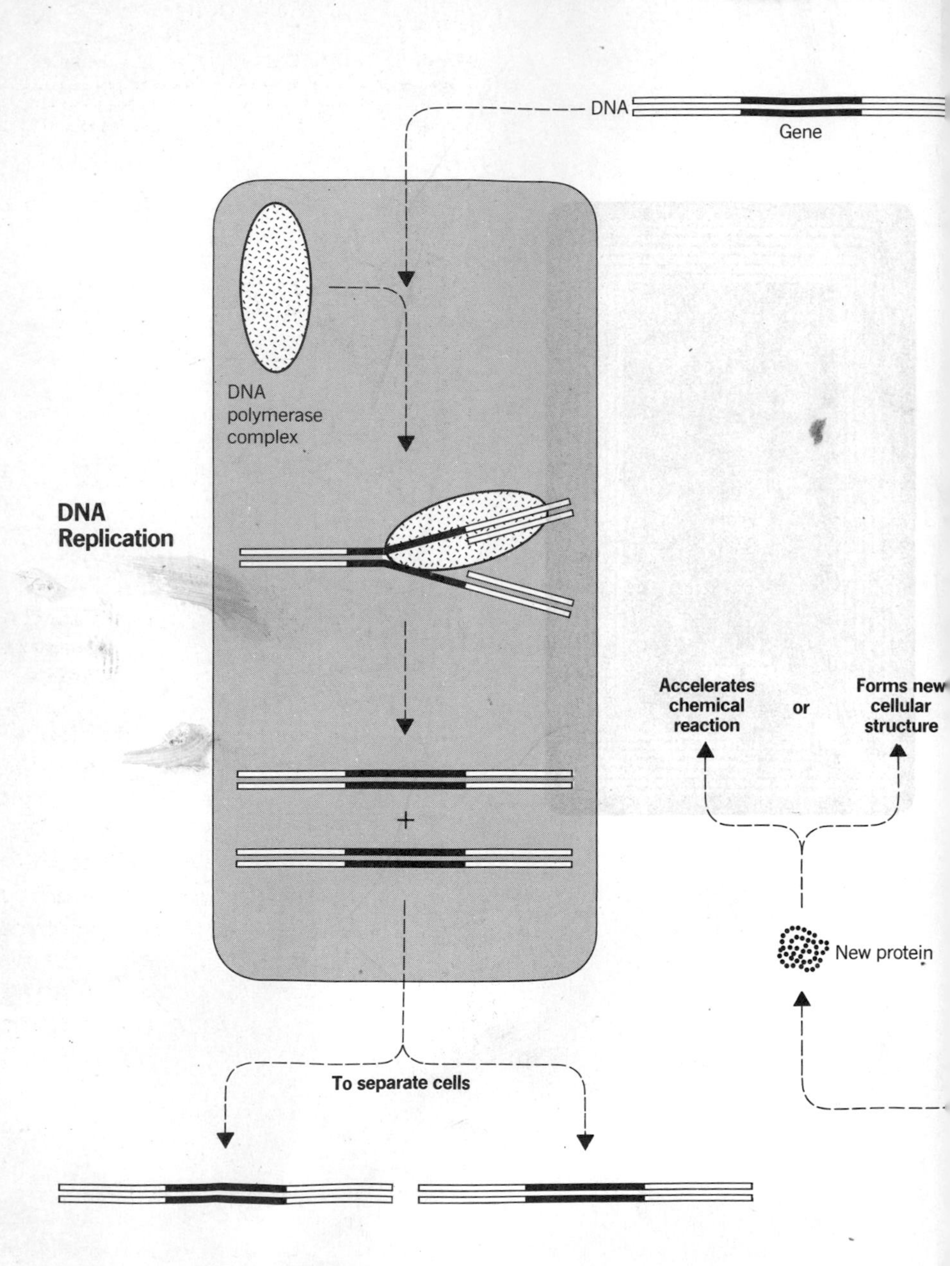